THE BACKYARD BUILDING BOOK

THE BACKYARD BUILDING BOOK

James E. Churchill

Illustrations by James E. Churchill II

 **Stackpole Books**

THE BACKYARD BUILDING BOOK

Copyright © 1976 by
James E. Churchill

Published by
STACKPOLE BOOKS
Cameron and Kelker Streets
P.O. Box 1831
Harrisburg, Pa. 17105

First printing, September 1976
Second printing, April 1977
Third printing, September 1977

*"Published simultaneously in Don Mills, Ontario, Canada
by Thomas Nelson & Sons, Ltd."*

Printed in the U.S.A.

Library of Congress Cataloging in Publication Data
Churchill, James E   1934-
  The backyard building book.

  Includes index.
  1. Building—Amateurs' manuals. 2. Dwellings—
Remodeling—Amateurs' manuals. 3. House construction—
Amateurs' manuals. I. Title.
TH148,C625   690'.8   76-17609
ISBN 0-8117-2105-1

*This book is dedicated to Jolain and Trapper Jim,
who have kept me building since they were born.*

CONTENTS

INTRODUCTION

Every homeowner wants to add to the comforts of his home and improve his property at the same time, but all too often today's sky-high building costs frustrate this ambition. The cost problem can be licked, however, and this book tells you how.

Do your kids need a place to play around the house without making a shambles of the furniture or risking their lives in the street? Then make your backyard a playground by building a Kentuck Kamp Fort, a Forest Ranger Fire Tower, an Observatory and Photography Darkroom, or a Mississippi River Barge (see chapter 8).

Planning on having outdoor barbecues this summer but don't have an outdoor fireplace yet? No need to hire a mason to build one. Chapter 9 not only tells how to install a barbecue pit in the backyard but also gives complete instructions for building a picnic table and benches to help you enjoy all that mouth-watering outdoor cooking.

Want to provide your overnight or weekend guests with luxurious accommodations at low cost? Then build them a geodesic dome, modified yurt, or screenhouse for them to stay in (see chapters 5 and 9) and an outdoor sauna bath to help them relax (see chapter 9).

Who wouldn't like to grow and enjoy his own vegetables the year round? It's easy when you follow the instructions for building a greenhouse in chapter 2. You can choose the type of greenhouse best suited to your level of building skill, from a simple window greenhouse you can attach to the side of your home to an A-frame greenhouse to a full-size, conventional greenhouse. Perhaps, after growing your first bumper crop of greenhouse vegetables by following the cultivation directions in chapter 2, you will wish to can some of them. But how to can without messing up the house? The answer is the easy-to-build harvest kitchen described in chapter 4. Not only does it simplify the task of canning but its bins provide plenty of space for storing root crops.

Thinking of starting your own cottage industry, one that will bring in badly needed supplementary income in these days of inflationary prices? The possibilities offered here are endless. Besides selling the produce you grow in your own greenhouse and the preserves you put up in your harvest kitchen, you can make furniture in your home workshop, lit by power provided by a wind generator made according to the instructions in chapter 6. The same chapter also tells how to fabricate a wind motor which will turn a water pump to keep a commercial fish pond open in winter or cool it in summer.

For each project described in this book the author has provided instructions sufficiently detailed to enable the homeowner with reasonable manual skills to complete it with maximum economy of time and effort. Hopefully, this publication will further the build-it-yourself trend so evident today among homeowners. City dwellers, suburbanites, and rural folks alike are finding that they save money and time and obtain a better structure if they do the work themselves.

A reader who builds all of the projects in this book should increase his estate by $50,000 or more at a cost to himself of from one-fourth to one-half that amount. This, of course, depends upon how shrewd a material buyer the reader is and where he lives. One good general tip on buying lumber is always to check more than one supplier since prices vary unbelievably.

The author does not intend to leave the reader without backup information. If at any time during the construction of any of the projects in this book a question is raised that cannot be answered by the book, just write him a letter and he will speedily make a report to you. Please enclose a stamped, addressed envelope. The author's address is:

James E. Churchill
Route #2, Box 160
Florence, Wisconsin 54121

With the help of this book you can proceed confidently in your program of home improvement, but do be careful of high places, sharp tools, and trying to do too much at one time.

Chapter 1

BASIC BUILDING TOOLS AND TECHNIQUES

The first section of this chapter tells how to use the tools which the amateur carpenter would expect to need for the projects covered in this book. Without knowing how to use tools, a person can scarcely expect to build anything. Fortunately all but a few tools are simple to use after a little practice. Maintaining them also is simple even though it is often overlooked. Possibly the most symbolic tool of the carpenter is the hammer.

HAMMER

A claw hammer, or a carpenter's hammer, is designed for driving and pulling nails. It is also very useful for dozens of other uses from cracking walnuts to ripping down an old building. Actually, two types of claw hammers are generally manufactured: the ripping hammer with fairly straight claws and the nail-pulling hammer with claws curved like an eagle's talon.

Each type of hammer is made with either a plain (flat) face or bell face head. The face is the part which strikes the nail. The bell face hammer is capable of driving the nail slightly below the surface of the wood without damaging the wood. The bell face is slightly harder to drive nails with, however, since the beginner might bend the nails by striking them off center. The plain face hammer has a broader surface for the nailhead to contact.

Actually, each can be used easily enough if the weight and balance are right. A good choice for the beginner is the 16-ounce size. From there you might progress to a 20-ounce size if you become a professional carpenter. Most carpenters want to sink a nail in four or five blows.

There is hardly anyone that hasn't started a nail with a hammer and driven it in the desired location. Still, most people do it wrong. The correct way to start a nail is to grasp it between the index and middle finger of your left hand. The point of the nail should be projecting from the back of the fingers. Place the point of the nail at the exact spot where you want it to be driven and give it a few light taps with the hammer, held near the end of the handle in the right hand. When the nail is well started, drive it the rest of the way in with solid blows, neither excessively hard nor too soft. The hammer hand must be kept approximately perpendicular to the nail to avoid bending it in either direction. Get used to this method and you will not show up at a party with a thumbnail blackened from smashing it with a blow from the hammer and your ability to drive nails accurately and swiftly will improve considerably.

Cheap hammers are available with cast iron heads. They are useful for hanging a picture or a few curtain rods but they shouldn't be considered for any serious carpenter work. By all means, when you buy a new hammer get one with a forged head and a good straight-grained hickory or steel handle.

Other hammers which the home craftsman will need to build the projects in this book are a tack hammer, mason's hammer, and half hatchet. A tack hammer is very useful for driving any nail less than one inch in length. It has a small head and handle. The mason's hammer is useful for breaking cement blocks, straightening forms, and chipping out unwanted masonry or brick. The half hatchet is used to chop rafters and the like.

SAWS

Both hand saw and power saws are frequently used in carpenter work. If you have electricity at your building site, then by all means use a power saw. It will make your work easier and more accurate. However, some projects just have to be done away from the power; therefore a hand saw must be used. A good hand saw in the hands of a skilled builder will cut with ease and accuracy.

Hand saws are generally classified as ripsaws or crosscut saws. The ripsaw has large teeth and is filed so the cutting edges of the teeth are shaped like miniature chisels. The teeth of a crosscut saw are small and are filed so they have cutting edges like a knife. A ripsaw is generally used to cut with the grain or "rip," while the crosscut is used to cut across the grain. Both saws have a "set" to the teeth so they will cut a "kerf" on the board. The kerf is wider than the blade of the saw and is useful to keep the blade from binding. The most useful lengths for sawblades are 24 inches and 26 inches. Probably the home craftsman should have a 24-inch crosscut saw and 26-inch ripsaw.

Making an accurate cut with a hand saw is difficult for the beginner because he works too hard at it. Do it like this: use a pencil and square to mark off the exact point of the cut. Pick up the saw in the right hand and lay the teeth on the waste side of the pencil mark so the pencil mark will not be cut off. Gently draw the saw back and push it forward at the correct place while holding it perpendicular to the cut with the left hand to start the cut or kerf. Continue guiding the angle of the sawblade with the left hand until the cut is about one inch deep. Then position the blade so it is about 45 degrees to the work if crosscutting or about 55 degrees if ripping. Keep your eye on the mark and draw the blade easily back and forth. Don't push down on either forward or backward strokes. Let the sawteeth do the work. If the work pinches it must be supported on the ends. When ripsawing, if the cut tends to push together, pinching the saw, keep it separated by pushing a small wedge or a screwdriver in the cut behind the saw. A little practice is prudent before the beginner cuts valuable lumber.

WOOD CHISELS

Wood chisels are made with a steel blade and a wood, plastic, or steel handle. The blade is beveled on one side and flat on the other to form a very useful tool for making joints, removing small amounts of wood from cuts, reshaping wood frames, and a thousand other tasks.

There are three general types of wood chisel: everlasting, tang, and socket. The everlasting chisel was designed to be driven with a steel hammer and it has a steel handle. It is used for heavy work such as chopping the end of a hardwood beam. The tang chisel gets its name from the tapered tang which is driven into a wooden handle. It is designed for hand work but will withstand light blows with a mallet. The socket chisel, so called because the handle fits in the socket of the blade, is made heavier and can be used either with a mallet or by hand.

All of these types of chisels are further divided into paring, butt, firmer, framing, mortise, and pocket chisels, depending upon the width and shape of the blade. Blade widths vary from 1/4 inch to 1 1/2 inches.

Actually, one firmer chisel with a socket handle and one 2 1/2-inch butt chisel will satisfy the needs of the home craftsman if they are kept well sharpened.

Before using the chisel on valuable wood get in some practice. Start by clamping a piece of soft wood in the vise so the grain of the wood runs lengthwise of the table. Grasp a firmer chisel in your right hand and place it bevel side down on the work so it will be pushed with the grain. Next, grasp the blade near the cutting edge end with the left hand. Now push forward on the blade so the chisel edge just takes a very thin shaving. Notice that by regulating the pressure on the blade with the left hand and the angle of the blade with the right hand the depth of the cuts can be regulated. When you become experienced at this, turn the chisel over and try cutting with the bevel side up. Notice that this takes a much deeper cut and, in fact, it must be driven with a mallet through most wood.

A wood chisel that is not sharp and clean is almost worthless. Therefore, the home craftsman must learn to sharpen his chisels to the correct angle almost as soon as he gets them or else he must be able to find someone to do this task.

Actually, learning to sharpen a chisel is simple enough, especially if it is kept in good enough condition so it only needs to be stoned. A firmer chisel is usually beveled to a 20-degree angle. When it is stoned, the cutting edge only has to be sharpened. Thus the chisel will be tilted at a 25- to 30-degree angle while sharpening. A few strokes are all that is necessary. When the bevel side of the blade is properly sharpened, the back side will have a wire edge. Turn the blade over and remove the wire edge by sliding the flat side of the blade on the stone a few times. It should shave the hair on the back of your hand.

The chisel sharpener is a device with a clamp and a roller. The chisel is clamped into this device and slid back and forth on the surface of the stone. It makes the job of sharpening foolproof since it maintains the proper angle.

If the cutting edge of your chisel is badly damaged, it must be ground on a power grinder. Do this by first setting the tool rest on the power grinder at a 25-degree angle. Check it with a protractor if you don't like to estimate the proper degrees. Then place the chisel blade against the tool rest so that the blade is square with the wheel. Move it back and forth over the rotating wheel to maintain an even cut. Grind out all of the nicks, remove the chisel from the grinder, and whet it to the proper hair-cutting edge on the whetstone. The chisel sharpener device mentioned previously can also be used to sharpen wood plane "irons" or blades.

PLANE

The plane is a very versatile and gratifying tool for the amateur to use. It will shave off the very finest of layers of stock to exactly the same depth every time, and it is almost impossible to do anything wrong with it if it is set up right.

There are several models of planes separated into categories according to the job they were designated to do. The block plane is the shortest standard plane, usually about six inches in length. The plane iron is mounted with the cutting edge bevel up, which is the opposite of the other standard planes. It gets its name from its common use of planing end grain, or "blocking in." Block planes are also useful for general planing where small amounts of material are to be used. The block plane can be used with one hand while the work is held with the other, which makes it more versatile.

The jack plane, so called because it is the "jack-ass" or workhorse of the planes, ranges from twelve to fifteen inches in length. The plane iron is mounted bevel down in this type, and a plane iron cap is mounted on the top of the plane iron close to the cutting edge to provide a sharper bend and greater shaving breaking action. It is used for smoothing edges and surfaces of boards with the grain. Generally, these two planes will satisfy the need of the home craftsman.

Other types of planes for specialized uses are the huge two-foot-long fore plane, the rabbet plane for cutting a step into a board, a plow plane for cutting groove work, a scrub plane for taking large slices of wood, and the circular plane for planing in concave surfaces.

A plane blade or "iron" is like a wood chisel. The cutting edge should be kept square with the sides of the iron. Just remove enough metal to take out any nicks or grooves. The plane has a built-in guide for grinding. Loosen the cap and set it back about 1/8 inch from the edge. Grind the bevel so it is about a 30-degree angle. When it is ground, remove the cap from the iron and clamp the iron in the chisel sharpener described in the chisel section. Stroke it a few times on the bevel edge and then reverse the blade and slip it across the stone a few times to remove the wire edge.

When you reclamp the iron to the cap, make sure it is only about 1/16 inch from the end for a minimum cut. More material can be taken with a large cut, of course.

With the description of saws, chisels, and planes we have covered many of the tools designed for cutting wood. However, before we start cutting wood we have to know where to cut. The next section of this chapter deals with measuring instruments, what they are and how to use them.

MEASURING INSTRUMENTS

Carpenter's Square

The picture that comes to most people's minds when they think of measuring instruments used in carpentry is probably the carpenter's square, or "framing" square. A framing square is so valuable that no home craftsman should be without one.

It is interesting how the steel framing square came into general usage. In the little town of South Shaftbury, Vermont a clear-thinking blacksmith named Silas Hawes wondered why a framing square couldn't be made from steel instead of hardwood as was the usual practice in those days. Wood framing squares were forever getting out of square as well as being clumsy to use. Besides, he was looking for some practical use for the pit saw blades he had lying around the shop. One day when work was slack he welded two of the saws together and stamped the scales on them. Pleased with his handiwork, he contacted a peddler who immediately placed an order for a dozen. The peddler soon came back with a huge order.

A few months later Hawes patented his steel square and opened a factory. In a few years he retired an independently wealthy man.

The framing square consists of the blade, which is 24 inches long and 2 inches wide, and the tongue, which is 16 inches long by 1 1/8 inches wide. The outside corner where the blade and tongue meet is called the "heel." The face of the square is the side where the manufacturer's name is stamped and the back is, of course, the side opposite the face.

The square is useful for straight line measuring and marking square corners. It is also a non-mechanized calculating machine which can be used to find board feet, length of rafters, and angles and degrees for various windows and stairways. In addition to all this, it functions as a table for fractional inch scales.

The usefulness of a good framing square perhaps can be best illustrated by using the problem of finding the proper length of a rafter for a 24-foot-wide building. It is desirable to have the roof rise 8 feet above the walls.

The first step in determining the length of the rafter is to find the pitch of the roof in relation to the width or span of the building. If the building is 24 feet wide, and the rise 8 feet above the wall, the pitch can be found by dividing 24 into 8, which is about .33, or 1/3. Thus the rafters will have a 1/3 pitch.

Now it will be necessary to find the rise in inches per foot of run of the rafter to be able to find the total length of the rafter. This is done by multiplying the pitch by "unit of span" or the constant, which is 24 inches. Thus the pitch, which is 1/3, multiplied by the unit of span, which is 24 inches, gives the result 8 inches. Hence, a 1/3 pitch roof on a 24-foot-wide building will have a rise per foot of 8 inches.

Now look on the face of the square for the words "length common rafters per foot of run." To the right of the words will be the inch marks. Find the 8-inch mark and look directly under it where the number 14.42 is stamped. This means that this type of rafter will have 14.42 inches of length for each 12 inches of span of the building. The rest is easy; just multiply the feet of span to the center of the building by 14.42. Thus a 24-foot-wide building will have a 12-foot span to its center, and 12 × 14.42 equals 173.04 inches. That is the total length of the span of the rafter. To find the proper cutoff marks, first *use the square to mark one end of the rafter,* so it will be exactly perpendicular to the edge of the rafter. Now measure off 173 inches along the rafter and make another mark. Use the square to project this line clear across the board also.

Now to find the cutoff marks, place the square on the sawn end of the rafter so the tongue is to your right and the heel points toward your body. Move the square until the 8-inch mark on the outer edge of the tongue is at the point where the sawn end and upper edge of the rafter is located. Then pivot the square so the 12-inch mark on the outer edge of the blade is located on the same edge of the rafter. Make a mark along the tongue across the width of the rafter. This marks the cutoff point for the rafter if no ridge board is used. If a ridge board is used, subtract ½ its width, which in most cases would be ¾ inch.

Now move the square to the heel mark and place it so the 12-inch mark on the body is located exactly where the measuring line and the upper edge of the rafters intersect. Then, as before, pivot the tongue so the 8-inch mark on the tongue is on the upper edge of the rafter. With the square in this position mark the heel and cut along the outer edge of the body of the square. If you cut the rafter off here, of course, there will be no tail on the rafter for use as an overhang, or eave. In actual practice a measuring line instead of the upper edge of the rafter is used. This is described under Laying Out the Pattern Rafter in chapter 4.

Ruler

Another useful measuring instrument that the home handyman frequently uses is the folding ruler or zigzag extension rule. This ruler, usually 6 feet long extended and 7½ inches long folded, has a brass slide-out rule in one end section for making inside measurements. Its major uses are for a distance marker or marking gauge. In addition to the 6-foot folding ruler most handymen have a steel tape rule, usually 12 feet long. For easy inside measuring the tape body is exactly 2 inches long. The end hook of quality steel tapes is loosely attached so that it will slide its own width in a tiny slot in the tape. This is necessary so both the inside and outside measurements are the same. Don't attempt to tighten the rivet holding the end hook to the tape since that would destroy the sliding action of this hook. Steel tapes are very useful for surface measurements but are hard to use for space measurements.

BRACE AND BIT

A tool which will be useful for building the **A** frame and some other projects in this book is the bit brace. Bit braces are usually widely available in hardware stores. They are the most useful of the hand-held tools for drilling large holes. They provide great turning leverage because the "sweep," or circle, in which the brace handle turns is relatively large. The sweep is usually about 10 inches and can be as high as 14 inches in some heavy-duty models. The bit brace can be used to drill holes of from ⅛ inch to about 2⅝ inches.

In a pinch the bit brace can be used to drill holes in metal also.

The type of drill used to drill holes in wood is called the auger bit. It comes in two general types: the single-thread form for making fast cuts in wet, gummy wood and the double-thread form for drilling in seasoned wood. The size of auger bits is indicated by a single number representing sixteenths of an inch, stamped on the square end of the shank. The number 12 means ¹²/₁₆, or ¾ inch.

Wood auger bits can easily be sharpened by the home craftsman. In fact, there is a special file made for this purpose called the auger-bit file. However, a small mill file is often used. When you sharpen an auger bit, a close examination will reveal that each cutting edge of the drill has a bevel like a chisel. Sharpen only the bevel, nothing else. Also try to keep the bevels or cutters even with each other as you sharpen. For a very keen edge, finish the job with slip stone.

FASTENERS

Some discussion of fasteners used to hold materials in their proper places is in order here. The most common fastener, of course, is the nail. Nails have been used in America since Colonial days. During that time they were all made from metal rods forged especially for the purpose. In fact, many a New England farmer supplemented his income by making nails for sale in pre-Revolutionary days. The struggle of the Colonies for mass production methods during the Revolutionary War brought about the invention of nail-making machines. Nail-making methods have

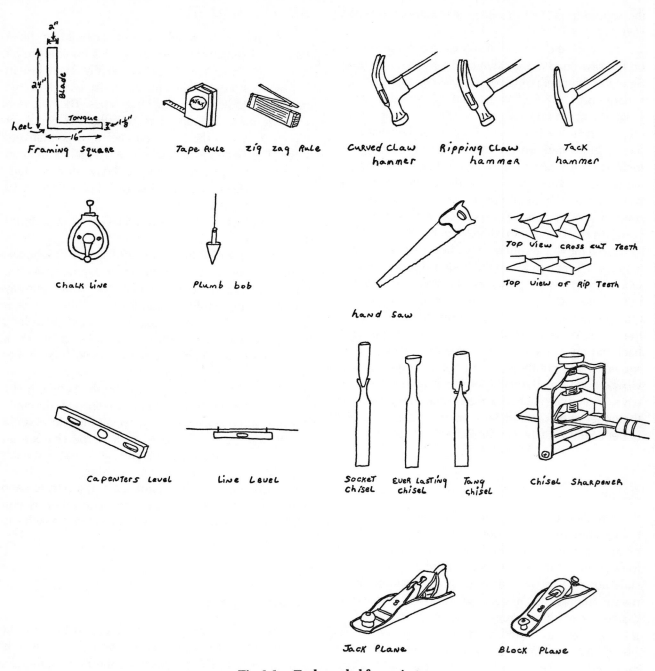

Fig. 1-1. Tools needed for projects.

been improved until now a modern nail-making machine can turn out about 500 nails per minute.

Nails are made in widely varying sizes and are generally designated by the penny system. The penny system of designating nail sizes came about, reportedly, because the nails were sold for so many pennies per hundred. Thus, six-penny nails sold for sixpence a hundred. If you have priced nails lately, you have noticed that you definitely can't buy 100 six-penny nails anymore for six cents. Common nails are made from two-penny size to sixty-penny size. The two-penny nail is 1 inch long and each size above that is ¼ inch longer. Thus, a six-penny nail is 2 inches long.

In the projects in this book both common nails and finishing nails will be used as well as spikes. A common nail has a thin shank and flat head. Finishing nails have a small round head and spikes are thicker in diameter than common nails although they may be the same length in some large sizes. Very long rafter spikes are also available and can be purchased at hardware stores by asking for them by name and designating the length desired.

Besides nails, wood screws and their application will be useful information to the homebuilder. Wood screws are manufactured from steel, iron, brass, copper, and bronze. Ordinarily, steel or brass screws are used in general carpentry work. They vary in length from ¼ inch to 6 inches and are designated by numbers 0 to 30. In the case of screws the number does not designate the length but the diameter. Many lengths are found in each numbered size, and it is important to know what length will give the best service when obtaining the screws. The right length will usually be determined by the width of the material being fastened. The full length of the threads of the screws should be buried in the supporting member. When all the threads are properly seated in the supporting member, a screw will give approximately twice as much holding power as a nail of the same diameter. For very heavy work lag screws should be used. These large wood screws are turned with a wrench and are widely used for fastening large timbers together. Most hardware stores have them.

Holes are drilled for placing wood screws. The first hole should be deep enough to contain the shank. It should be nearly as wide as the shank. The second hole should be drilled at the bottom of the shank hole to approximately half the depth of the threads on the screws. It should be about half the diameter of the threads at the shank also. The exact diameter of this second hole will be determined by the hardness of the wood. In soft wood with small screws no second hole at all need be drilled. In very hard or brittle wood the diameter of the second hole should be larger than half the diameter of the threads.

Screws are driven by screwdrivers and every home craftsman needs an assortment of these. Screwdrivers are made to match screw sizes. Thus, five different-size screwdrivers cover the complete range of screws. The largest-width blade for a screwdriver is ½ inch; the smallest, ¼ inch. The largest takes size #20

to #24 screws; the smallest, #6 to #8 screws. Screws with a phillips head can all be turned by four different-sized screwdrivers. Many times screwdriver sets are sold to cover the full range. A screwdriver fits a screw when the width of the blade is as wide as the screw and the thickness of the blade just fits the slot.

Other fasteners that will be needed are machine bolts. They are made in sizes from ¼ inch diameter to 1¼ inches in diameter. Probably the sizes most used in building wooden structures are ⅜, ½, and ⅝ inches. This refers to the diameter. The heads of bolts most often used in building are called carriage bolt heads. This type of bolt, when inserted in a hole, will grip the wood and not turn when the nut is screwed on it, because of a square section under the top of the head.

Generally, "open end" or adjustable wrenches are used to turn the nuts or bolts in building construction. The proper size wrench for standard bolts can be calculated by multiplying the diameter of the bolt by 1½.

For example, if you wish to find the proper wrench size for a ½-inch diameter bolt, multiply $1\frac{1}{2} \times \frac{1}{2} = \frac{3}{2} \times \frac{1}{2} = \frac{3}{4}$. Thus, a ¾-inch wrench will fit the head of a ½-inch bolt.

Other than very specialized kinds of fasteners, nails, screws, and bolts hold almost all buildings together.

PLUMB BOB, CHALK LINE, AND LEVELS

The plumb bob, chalk line, and level are used so often in carpenter work that it would hardly be possible without them.

The plumb bob consists of a pointed weight attached to a string. It is used to find a point perpendicular to the horizon, or "straight up and down." A commercial type of plumb bob often used is called the adjustable plumb bob. It consists of a pointed weight called a "bob," a friction reel which contains the string, and a suspension ring which is attached to the end of the line. It is handy to use because any desirable length of line can be pulled out of it by jerking on the suspension ring.

Plumb bobs are useful for positioning the center of a rafter peak over the center of the upstairs floor, making sure walls are straight up and down, aligning posts, and dozens of other uses.

The carpenter's level is used not only to find a straight up-and-down line but also to find the lines in the same plane as the horizon, or "level." It is used to guide the placement of boards and to check the construction. It would be extremely hard to build a structurally sound building without a carpenter's level.

Usually made of wood, although aluminum and steel are also used for the body, the carpenter's level consists of a body containing at least two glass tubes. These tubes are filled with a non-freezable liquid, and they each contain a bubble. One tube is used to check "level" and is placed parallel with the edge of the level. The other tube is placed perpendicular to the edge of the level and is used like the plumb bob to check "plumb."

Another level which is used for the projects in this book is the line level. A line level is a small metal tube containing a glass tube filled with a bubble-containing liquid. It is hung from a string and indicates when the string is level so that the string can be used as a guide in placing foundations and posts.

A chalk line is often used to hang the line level from. It consists of stout string often contained in a reel, and is called chalk line because a frequent use is to snap a line on a floor or other surface. This is done by coating the string with special chalk, usually blue chalk. Then the line is pulled tight between two points and held there. Next, the line is lifted to make it tighter still and then released suddenly so that it slaps the floor and leaves a chalk mark. If this is done right, the line will be straight and true between two points.

The foregoing information covers most of the tools used in day-to-day work by the home craftsman in general and by the builders of the projects in this book in particular. Now for some discussion of electric wiring, plumbing, and painting.

PLUMBING

Plumbing for a house in the outback consists of pipes to carry the water into the house from the well, tanks to contain a supply of hot and cold water so it is ready for immediate use, pipes to distribute hot and cold water to the proper place, and pipes which carry the waste water and effluent from the house to a sewer or septic system.

The first item to consider for a builder of a new home is how to get the water from the well to the house. In most cases this requires an electric pump. If the well is not sandy, a submersible pump is suspended in the water of the well and pumps the water up to the cold water storage tank in the house. It is very efficient since it does not have to draw water but merely push it along. Submersible pumps are usually wired to be activated by a pressure-sensitive switch on the cold water tank. When the pressure in the tank drops low enough the switch activates the pump, which pumps enough water to bring the pressure up to a predetermined setting. This then causes the pressure switch to open and stop the pump.

Submersible pumps, however, are subject to damage from sandy wells and cannot be used there. In sandy wells a jet pump is a good choice. Jet pumps are located above the well and draw water from it by means of two pipes connected to a "jet." They are simple to install, easy to service, and in most cases long-lived. They are not quite as efficient as submersible pumps and they must be located in a separate well pit or in the basement of the house if the well is close enough and low enough. Jet pumps also are activated by pressure switches.

The piston pump is the other widely used house water system pump. It can be used with shallow wells only and it often is a combination of pump and storage tank. The disadvantages of the piston pump are its noise and relative inefficiency. It is economical and very simple to install, however.

All of the water pumps described will do a good job in delivering water to a storage tank. The cold water storage tank comes in a great variety of capacities, but the larger the tank that is used the less often the pump has to be activated, which increases its life and makes more water available for emergencies. A large tank of water, however, is very heavy and the builder should be certain his floor is strong enough to support it.

When the builder starts to think about the location for the plumbing, he should group everything as close as possible to the cold water tank. Thus, the hot water tank and the various sinks and water closets should be close to each other if possible.

A bonus from this will be the simplification of

the drainage system. The drainpipes from an independent septic system all empty into the soil pipe. The soil pipe is a four-inch diameter pipe that extends from the sewer line in the soil through the roof. This ventilates the sewer system and all of the installations inside the house. The drain lines from the toilet, bath, and sinks all go to this main pipe. For simplification of plumbing have the kitchen and bathroom as close together as feasible, especially when the building does not have a basement.

Every conceivable type of plumbing problem is dealt with in publications these days, and the prospective builder should probably make a trip to the public library to consult a few texts on plumbing while he is installing his water system. The use of plastic pipe will also greatly simplify the installation of plumbing and it should present no problem to the self-reliant readers of this book.

WIRING

Generally speaking, the installation of home wiring must be done by a licensed electrician. However, once the "juice" is in, which means that the main electrical junction box is wired in by an electrician and the main circuits run, the home electrician can install whatever other circuits that he desires.

ROOFING

Probably of more concern to the homebuilder will be the installation of the roofing. All of the projects in this book have roofs that can be installed by the home handyman with a minimum of equipment. Perhaps the most difficult of the roofs to cover is the A-frame since it is so steep. In fact, the pitch of the roof has much to do with the way the roofing is applied and what roofing is used.

The pitch of a roof is the amount that the roof rises in a foot. A roof that rises 6 inches in 12 inches will have a $\frac{1}{4}$ pitch. A roof that rises 8 inches in 12 inches has a $\frac{1}{3}$ pitch, and a roof that rises 12 inches in 12 inches has a $\frac{1}{2}$ pitch. This can be checked with the framing square if desirable, but an estimation can also be easily made.

Wood and asphalt shingles can be used with a roof that has a minimum of a 4-inch rise in 12 inches. Mineral-surfaced roll roofing can be applied to a roof

with a rise as little as 2 inches in 12 inches. Flat roofs are usually covered with felt and hot tar layers; professional roofers usually apply four layers and then cover the top layer with crushed stone.

New construction should be covered with #15 or #30 felt and then have the shingles applied on top of the felt. The felt is applied horizontally, starting from the bottom of the roof, and the layers are overlapped from 2 to 6 inches. The felt is nailed down with a $\frac{1}{2}$-inch roofing nail every 6 inches or less. Actually, the fewer nails that are applied to the felt, the better, unless the felt will be left on the roof long enough to be exposed to high winds before the shingles are applied.

Shingles are applied with a definite "exposure"—that is, the proportion of each shingle that is exposed to the weather. The steeper the pitch of the roof, the more shingle can be exposed. Manufacturers of asphalt shingles specify the proper exposure. Roll roofing has a color-coded exposure. Wooden shingles are made in three lengths: 15 inches, 18 inches, and 24 inches. Roofs with a 5-inch in 12-inch pitch have 5-inch exposure for the 16-inch shingles, $5\frac{1}{2}$-inch for the 18-inch, and $7\frac{1}{2}$-inch for the 24-inch shingles. Roofs having less than 5 inches of pitch in 12 inches have approximately 2 inches less exposure for each size shingle.

Shingles are often sold by the "bundle." A bundle is $\frac{1}{4}$ of a square. A square is 100 square feet. The roofer always estimates the surface of the roof by multiplying its width by its length and dividing by 100. This gives him the total squares of roofing needed. In addition, the starter strips along each roof edge are estimated by dividing the width of a shingle by the total feet of the roof edge. Asphalt shingles are usually 3 feet wide.

Start shingling the roof after the felt is applied by installing the starter strip at the edge of the roof. This starter strip is applied to make sure the rain runs over the edge of the roof and does not soak the edge of the roof boards. A starter strip is usually a row of shingles placed in the reverse of the normal position so that the solid edge projects down. To keep this starter strip straight a chalk line is usually snapped along the roof where the upper edge of the shingles will fall.

After the starter strip the first layer of shingles is nailed in place. Many roofers start at the center of the roof and work outward, cutting off the projecting edges of the shingles where they extend over the end. A chalk line is used to keep this first row of shingles straight also.

Each asphalt shingle is nailed in six places with rustproof roofing nails, ⁵/₈ inches long. When the peak of the roof is reached, special peak shingles are installed, or a metal roof ridge can be used. Flashing, thin metal sheets sold in hardware stores, should be installed around chimneys. It is usually copper or aluminum. It is nailed in position to make sure there is no leak between the chimney and the last row of asphalt shingles.

One of the projects in this book, the A-frame cabin, has to have a "chimney cricket." A chimney cricket is a small peaked ridge built behind the chimney. The cricket is as high as half the width of the chimney. In the A-frame where insulated pipe is used the cricket should be 18 inches high. A cricket extends from the chimney on the level to the roof. Sides to the roof are then built to carry away the rain. The cricket is covered with shingles like the roof.

Wood shingle application is similar to asphalt. First a layer of 15-pound felt is applied. Then the first row of shingles, the starter course, is applied. It extends over the edges of the roof ¹/₂ to ⁵/₈ inch.

When the first course is laid over the starter course, line the butts up with a straight 1 × 4. Stagger the joints between the shingles so they are at least 1 inch apart. Nail each shingle with two five-penny rustproof nails placed approximately in the horizontal center of the shingle. When the peak of the roof is reached, double the courses or use a 1 × 4 and a 1 × 6 ridge seal. When the 1 × 4 and 1 × 6 are used, nail the 1 × 6 along the roof so it extends over the peak ³/₄ inch. Then butt the 1 × 4 against it and nail it in place. Use eight-penny nails. Further, nail through the joint between the boards to form a tight joint. As an added precaution, many homebuilders seal the joint with asphalt. It, of course, should be painted to complement the roof shades.

Roll roofing makes a very good roof and it is probably the simplest and least complicated to apply. First the roofing should be rolled out and allowed to lie long enough so the wrinkles are smoothed out, especially if the weather is cool. Roll roofing is cemented to the roof with asphalt cement. No nails need be used unless the roofing starts to slip as on very hot days.

Start the first course by snapping a chalk line 35 inches up the roof from the edge. This allows the roofing to project over the edge about an inch. Cover the entire area of the roof under the first course with a layer of asphalt cement. Then roll the first layer of roofing in place. No felt need be used with roll roofing.

Roll roofing is lapped 18 inches. This is indicated by the way the roofing is made. One-half of it is covered with mineral surfacing, the other half is smooth. When the first layer of roofing is applied over the starter course, the upper half of the first course as well as the 19 inches of roof above the starter course is covered with cement. Then the first layer of roofing is rolled out over it. Each succeeding layer is also applied that way. Each layer of roofing should be rolled to bond it to the layer underneath. If a roller isn't available, the roofing should be walked down so it will bond. When the peak of the roof is reached, apply a narrow course of roll roofing right over the ridge, apply a commercial ridge cap, or use the 1 × 4 and 1 × 6 method mentioned above.

Chapter 2

GREENHOUSES AND HOTBEDS

Last January I visited a friend's greenhouse when the temperature was balanced precariously at the last plus number on the thermometer and an arctic wind swirled sheets of powdery snow across frigid fields. When he opened the door, tropical air enveloped me and I stepped forward into an oasis of tender green leaves and bright blossoms. It was like stepping from the Arctic Circle to a South Sea island in one step, and the warm, snug sensation this produced will remain in my mind as long as memories have meaning.

My friend raised flowers primarily for his own enjoyment but he sold some also. As a result, his greenhouse wasn't a liability financially. In fact, he said, if he chose he could make a good living raising plants for the various retail outlets. In the months that followed I looked further into the possibilities of a greenhouse and found that indeed there is hardly a cottage industry that has better potential for the green-thumber than raising house plants and food

plants in the greenhouse for the various retail and wholesale outlets.

Moreover, if you believe as I do that the world food supply has peaked and except for peculiar good years we are going to face continuing food shortages from now on, then the construction and use of a greenhouse takes of an urgent tone. Perhaps the day will come when every home will be built with an attached greenhouse as a matter of course, and raising vegetables and fruits the year around for the family food supply will be a required course at public schools.

A few years back all greenhouses were made of glass with metal or wood frames. They came in two general classifications: lean-to and freestanding. The lean-to greenhouse was the most economical to construct because it was built against a house or garage wall. Usually heat from the house was used to heat the lean-to greenhouse and, since it could be entered from the house at any time in any weather, it was the popular choice of hobby operators. Some disadvantages were that it could hardly be expanded when, as usually happens, more growing space was eventually desired and diseases, humidity, and cleanliness were hard to control.

Freestanding houses could be made to any size but they required a separate heat supply and expensive plumbing and wiring, and were far beyond the financial means of most homeowners, since they cost several thousands of dollars.

Fortunately most of these problems have now been met with the development of flexible fiberglass and clear plastics. Today anyone with a few simple tools can build a greenhouse with a wooden frame and cover it with plastic; depending on the size, it can cost less than a new suit.

Also, flexible fiberglass or plastic can be applied much tighter at the seams and is a far better insulator than glass. As a result, climate control is made easier. A further aid is the recent practice of installing a second layer of clear plastic on the inside of the greenhouse frame. This creates a dead air space which is a very effective insulator without sacrificing light transmission. In fact, window greenhouses made with double walls actually aid the home heating plant since they receive more solar heat then they lose, a process unlike the continuous rapid conduction of heat through glass that usually takes place when windows are installed.

WINDOW GREENHOUSE

Window greenhouses are easy to construct and the number of plants that can be raised in one is truly amazing.

A dear, old friend of ours, though born and raised in the country, was forced in his declining

Fig. 2-1.   Window greenhouse.

years to move into a large city. Restless and out of sorts without plants, he finally persuaded his landlord to let him install a window greenhouse on his south-facing, sixth-floor apartment window. In this growing space he managed to raise radishes, lettuce, and even two tomato plants. His crowning glory, though, was two miniature cornstalks which produced two ears of corn for his Christmas dinner.

The first year we went to live in the woods we arrived at our none-too-tight log cabin about the first of September. In Wisconsin that is the start of autumn. One of the first things we did was to enclose the three windows on the south side of the cabin with a fiberglass and 2 x 4 platform. Before the ground froze we managed to get enough leafmold, peat moss, and loam together to fill two dozen large tin cans. This was our garden and we feasted all winter on greens and radishes which grew from these cans. Our only source of heat was a wood stove but we didn't lose one single plant, probably because we covered them on very cold nights with strips of an old wool mackinaw. Once we enclosed a screened porch with clear plastic and managed to raise dandelion and mustard greens all winter in it with hardly any additional heat.

Commercial models of window greenhouses are available, of course, but anyone can construct his own. Perhaps the first consideration will be what window or windows to use. They should face south, southeast, or southwest unless you intend to supply supplemental lighting. In that case any exposure will be all right.

Generally speaking, turning windows into a window greenhouse sacrifices the view from the location. Thus, if you have an unobstructed view of a snowcapped mountain or a nude beach, you may want to reconsider the location.

MATERIALS LIST FOR WINDOW GREENHOUSE

1. 2 8-foot 2 × 2's
2. 2 8½-foot 2 × 2's
3. 2 3-foot 2 × 2's
4. 2 2-foot 2 × 2's
5. 2 1-foot 2 × 2's
6. 9 4-foot 2 × 2's
7. 12 ¼ × 3½-inch cadmium-plated bolts
8. ½ pound six-penny nails
9. ½ × 48 × 36-inch hardboard shelf
10. ½ × 48 × 24-inch hardboard shelf
11. ½ × 48 × 12-inch hardboard shelf
12. 24 #6 × 1½-inch flathead wood screws
13. Paint and wood preservative
14. 12 ten-penny duplex head nails
15. 4 × 25-foot roll .037 translucent fiberglass
16. 100 feet ¾-inch screen bead
17. Box of brads
18. Roll of fiberglass wrapping insulation, 3 inches wide

The first step in construction is to procure the lumber. See the materials list for the amount and size. The 2 × 2's should be knot-free redwood or the equivalent. Saw out the desired sizes and then plane or sand a radius on all corners of all the framing members. This is to remove all corners and wood splinters so the covering won't be damaged. Next add two coats of a good preservative. Mix you own or procure a commercial mix. Let it dry overnight.

Next morning temporarily tack the structure together with 6D finishing nails. Now check Figure 2-2 carefully to see that all parts are in the right position and then drill ¼-inch holes through the members at the joints to install the ¼ × 3½-inch cadmium-plated carriage bolts, nuts, and washers, one at each joint. Next, place the ½-inch hardboard shelves in their proper position and use #6 × 1½-inch wood screws to fasten them in place. Countersink the heads.

The final step is to paint your creation with colors that harmonize with the exterior of the house. The interior surfaces can be painted to harmonize with the interior of the house, but make sure they are a pale color to reflect as much light as possible—white, of course, being the best of all.

The window greenhouse is fastened on the side of the house with duplex framing nails or wood screws. However, before it is installed, the insulation should be put in place. First place the window greenhouse in the desired position on the side of the house and make a pencil mark all around it. Set it aside and staple or tack strips of fiberglass insulation inside the pencil mark. The insulation sold for wrapping water pipes works well; of course, if scrap insulation is available, it can be utilized also.

Once the insulation is installed the window greenhouse is ready to be put in place, but before installing it remove the window's upper and lower sash. Generally they are held in place with slider strips and

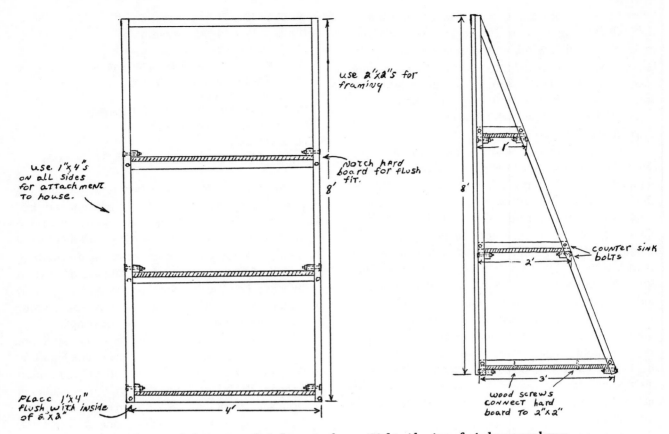

use 2"x2"s for framing

use 1"x4"s on all sides for attachment to house.

Notch hard board for flush fit.

8'

Place 1'x4" flush with inside of 2"x2"

4'

8'

1'

2'

3'

counter sink bolts

wood screws connect hard board to 2"x2"

Fig. 2-2.   *Left:* front view of window greenhouse. *Right:* side view of window greenhouse.

can be pried loose with a wide flat smooth pry bar. When the windows are removed, store them in a safe place where they can be stood up. Laying them down will almost certainly cause the panes to warp if they aren't supported some other way. Be sure to remember where you put them in case the cover of the greenhouse is damaged by a falling icicle or an inaccurately aimed football, and you have to reinstall the window.

After the window greenhouse is in position the plastic or fiberglass covering can be installed. Coverings range all the way from the very economical to expensive, with four-mill vinyl plastic probably being the cheapest material which will give satisfactory service. The recommended covering is translucent fiberglass in .037 thickness. It costs less than fifty cents a square foot at this writing. Rigid fiberglass panels are available also, and for a permanent installation consideration should be given to their use.

In cold climates a double wall covering should

be installed. In this case the inside covering will ideally be four-mill vinyl plastic since it will last for a long time if used in conjunction with translucent fiberglass, which redirects the sunlight. Make sure screen brads or other wood strips are used over a double thickness of the covering where it is fastened. Also make sure the covering is pulled tight to prevent the wind from rippling it, which will eventually wear it out. With this done, the window greenhouse is ready to use.

Now, it is possible that you will have windows in your house that won't be compatible with this size and shape of window greenhouse—for example, a horizontal window. In that case carefully measure your window, draw up your own plans, and proceed, using the same materials recommended for this type of greenhouse. At any rate, when you have your window greenhouse done, you can start gardening.

It might be surprising to note that we have a total of twenty-four square feet of growing area within our window greenhouse, with each shelf placed to receive full sunlight. This means that we can grow at least two dozen house plants, 1 dozen tomato plants, or a considerable collection of other vegetables.

The first step is to find or purchase containers to raise your plants in. Clay flower pots are fine as are coffee cans, wooden boxes, or the trays made especially for this purpose and sold by garden supply stores.

The soil for raising vegetables can be home-mixed by combining one part garden mold, two parts leafmold, one part well-rotted cow manure, and one part clean, sharp sand. Mix the ingredients very well and keep moist.

Lettuce, radishes, tomatoes, onions, and carrots are some of the food plants that can be raised in this planting space. Soak all seeds overnight before planting them directly in the trays where they will eventually grow. Tomatoes can be started in peat pots if desired and finally transferred to 8-inch pots. It is recommended the first shelf be used for tomatoes, since they will have to be provided with about two square feet of growing space to fruit properly. Smaller plants such as radishes and lettuce can be planted between the tomato plants so the space will not be wasted. When the tomatoes blossom, they must be hand-pollinated or they will not set fruit, since there are no bees or other insects to do this naturally. Pollinating is usually done with a small artist's brush. Simply stroke each blossom containing pollen gently with the brush. This should transfer enough pollen between the blossoms for fertilization.

After the tomatoes are grown to the ripening stage the plants should be turned so each fruit receives as much sunlight as possible. Further, if the foliage is excessively heavy, it should be trimmed back or tied up so it doesn't shade the ripening fruits.

All garden plants should receive an application of a good complete liquid fertilizer every two or three weeks. Determine the frequency by the growth of the plants. When they slow down or stop growing, apply fertilizer. However, if they don't respond to fertilizer, discontinue its use. Water frequently enough to keep the soil moist.

Lettuce should be ready to eat in five weeks, radishes in three. So make successive plantings to keep a steady supply available. Some people raise tomato plants on a successive basis also so that one or

two plants are bearing at any given time. Carrots have to have at least 6 inches of soil to bear usable roots. Use a small variety for best results (this is true for most vegetables).

Window greenhouses can be a source of never ending pleasure and interest to the plant lover. They can be made almost any shape that you desire to conform to the style of your house. Some people get so interested in "windowsill" gardening that they raise vegetables in one window, herbs in another, and flowers in still another. At the very least this is an economical way to see if you are compatible enough with indoor gardening to want to go to a hotbed or even to a larger, more expensive greenhouse.

HOTBED GARDENING

The next progressive step in winter gardening after the window greenhouse may be the hotbed. Now, a hotbed appears to be a form placed on the ground and covered with transparent material. There is actually much more to it.

First select a site where the drainage is very good. Directly on the top of a sand knoll is almost ideal but many locations are satisfactory. Stake out and excavate a rectangle about eight feet long and four feet wide to a depth of about two feet. Fill this rectangle with horse, cow, rabbit, or goat manure. If no manure is available, fill it with sawdust, wood chips, hay, green grass, or other organic matter that will decompose and generate some heat while it does. When the excavation is full, put on an old pair of boots and pack the manure so that there is about six inches of space between the manure and the top of the ground. On top of the manure place about 2 inches of moisture-absorbing sand.

To make our hotbed a reliably heated enclosure for raising plants requires the addition of an electric heating cable, or heating tape, which is placed on the sand. Use the type cable made especially for soil heating. This cable is manufactured with a built-in thermostat that keeps the temperature about 70 degrees for optimum plant growth. It will require a cable 160 feet long to protect the 32 square feet of growing area enclosed within this hotbed. Make sure it is placed so as to produce even heating and also make sure the thermostat is located near what will probably be the coldest corner. This type of cable

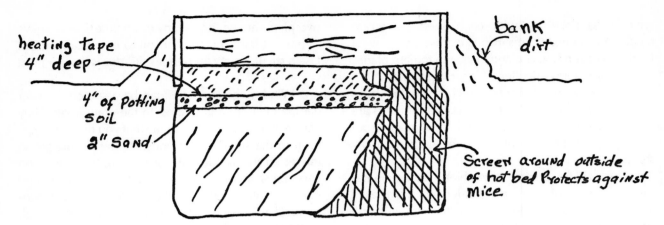

heating tape
4" deep

bank
dirt

4" of potting
soil

2" sand

Screen around outside
of hotbed protects against
mice

Fig. 2-3.   Cross section of hotbed.

shuts off automatically when the heat gets too high, and, depending on how much heat is generated by the decomposing organic matter, it might use very little energy. When the cable is in place, lay a screen or fine-mesh chicken wire over it and all around the perimeter of the hole. This prevents digging tools from damaging the cable and it also keeps rats and mice from setting up living quarters in your nice warm hotbed.

The final step is to finish filling the excavation with compost or garden soil made by the same formula as we used for vegetables in the window greenhouse. Fill the excavation about two inches over the top to allow for settling. Keep the connector plug of the soil heating cable from being buried, as it will be connected to the house current with an extension cord.

Now, with our "ground work" done we can install the frame. Procure and cut 1 × 12-in redwood or cedar boards to the lengths in Figure 2-4. Pine or fir can be used too if it is treated with a good preservative. Nail the boards at the corners with 2 × 2-inch cleats; use 10D plated nails. When the frame is completed place it over the excavation and level it up with bricks or similar material. Then stake it down on all four corners and shovel soil against the sides of the frame to form a bank to within an inch of the top. Now install two inches of foam insulation completely around the interior of the hotbed wooden frame. Two inches of foam is equivalent to six inches of fiberglass insulation for this purpose. Use white foam so it will

reflect light as much as possible. If you must use fiberglass, place the reflecting surface inside.

When the foregoing is all complete we can make the cover. It is generally expedient to make a double wall cover for the hotbed. Experiments have shown that up to twenty degrees lower temperature can be tolerated by plants in this hotbed if the cover is double wall plastic. Make the cover frame of 2 × 4 stock as shown in Figure 2-5. Put a cross brace of 2 × 4-inch or 2 × 2-inch in the center. Miter the corners at a 45-degree angle and fasten the joints with metal joint nails available in hardware stores and lumber yards. The plastic or fiberglass covering is stapled or nailed and then wrapped around the corners. Screen door hooks and eyes can be utilized for holding the cover in place. This is necessary to prevent a high wind from blowing it off. If desired, one side of the frame can be hinged. This will allow the cover to be raised slightly for ventilation, which should be provided about once a day, except in below zero weather. Don't raise the top at all in very cold weather as the plants are likely to be set back or made sterile by being chilled. On very cold nights, cover the hotbed with hay, old rugs, insulation, or other material to prevent heat loss.

This hotbed is effective for raising cool weather plants such as lettuce, endive, cabbage, broccoli, spinach, radishes, parsley, beets, and all types of pot greens, such as chicory. It is also very effective for starting warm weather plants such as tomatoes, peppers, cucumbers, squash, and melons. These plants

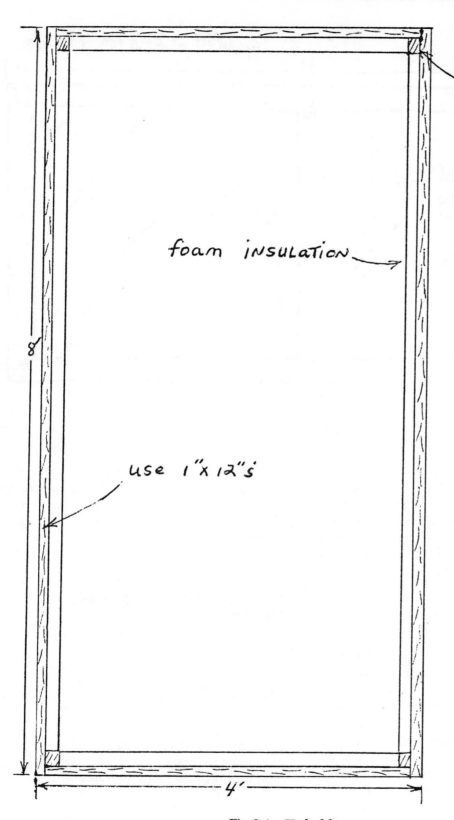

use cleats to
join boards
together

foam insulation →

use 1"x 12"s

8'

4'

Fig. 2-4. Hotbed frame.

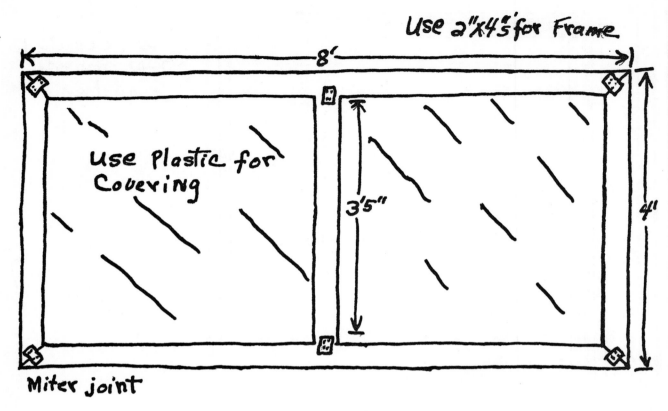

Fig. 2-5. Hotbed cover.

can be hardened gradually as spring comes by reducing the intervals of heat. Finally, when the heat is not used any longer and the top can be left off the frame without damage, the started plants can be transplanted to your garden. Also, such plants can be started in this hotbed and then later set out in the greenhouse to continue growing.

A-FRAME GREENHOUSE

The A-frame greenhouse costs very little to build or operate, yet works well for growing early varieties of tomatoes, peppers, melons, and many other plants.

This type of greenhouse is almost the easiest structure to build that it is possible to design. Even if you have never built anything in your life, but know what a hammer, nails, and boards are, you can build this greenhouse. What's more, it has many advantages over conventional buildings. For instance, snow and rain will run off this structure so fast that no leakage or snow load problems are likely to be encountered. The location for this type greenhouse isn't

too critical either, since light will penetrate every corner because of its walls being perpendicular to the sun's angle. Another very important advantage is that no foundation is necessary since the weight is minimal.

MATERIALS LIST FOR A-FRAME GREENHOUSE

1. 20 $\frac{1}{4}$ ×6-inch carriage bolts, nuts, washers
2. 8 10-foot 2 × 4's or poles
3. 2 12-foot 1 × 10 boards
4. 15 4-foot 2 × 4's or 3-inch poles
5. 2 2-foot 2 × 4's or 3-inch poles
6. 1 24 × 60-inch door, hinges, lockset
7. 2 6-foot 2 × 4's
8. 1 3-foot 2 × 4 for door frame
9. 18 feet of door stop
10. 4 1 × 12 gussets for 2 × 4 A-frames
11. 3 rolls of 4 × 25-foot .037 fiberglass
12. 100 feet of 1 × 2 furring strips
13. 300 square feet of polyethylene for inside cover

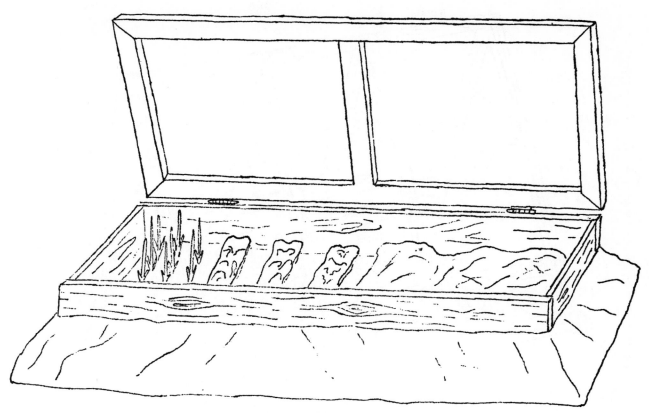

Fig. 2-6. Hotbed garden.

One of the first steps in building is to acquire poles for the rafters. Cedar is good as is redwood and treated fir or pine. Knot-free 2 × 4's also can be used if desired. The eight vertical rafters should be 10 feet long. Poles should be 4 inches in diameter on the small end, which of course, is always used at the top. They are notched at the top to overlap each other. The joints are held together with 1/4 × 6-inch carriage bolts or nailed together with 20D spikes. If 2 × 4's are used, they can be mitered and fastened with joint nails.

Two feet under each joint a cross brace is installed. This also can be either a 2 × 4 or a pole, and it likewise is notched into the vertical rafters. The cross braces can be nailed with suitable spikes.

Horizontal rafters are also used. They can be spaced at 3-foot intervals and notched into the vertical rafters. Use nails or spikes to fasten them also. Try to keep them flush with the outside surface of the vertical rafters.

However, before fastening the side rafters make up all four frames for this 12-foot greenhouse. Note that the front frame has the door frame installed. The two center frames do not have a cross member at the bottom while each end frame does. After the frames are made up, coat them with a good preservative and set them aside while you get the specialized floor ready for this greenhouse.

Much heat loss can be prevented and even some heat generated by the following method of designing the floor. First, lay out a trench one foot wide all around the perimeter of the prospective greenhouse site. The outside wall of the trench should form a rectangle six feet wide and eleven feet long. Remember to orient the greenhouse east to west if possible. Also try not to place it under a tree where falling limbs are likely to damage the covering. However, it is wise to take advantage of any natural or man-made wind breaks, since wind blowing directly on the greenhouse can cause noticeable heat losses. See the section Full-Size Conventional Frame Greenhouse later in this chapter for squaring up the structure properly with other buildings or a road.

When the trench is properly located, excavate it

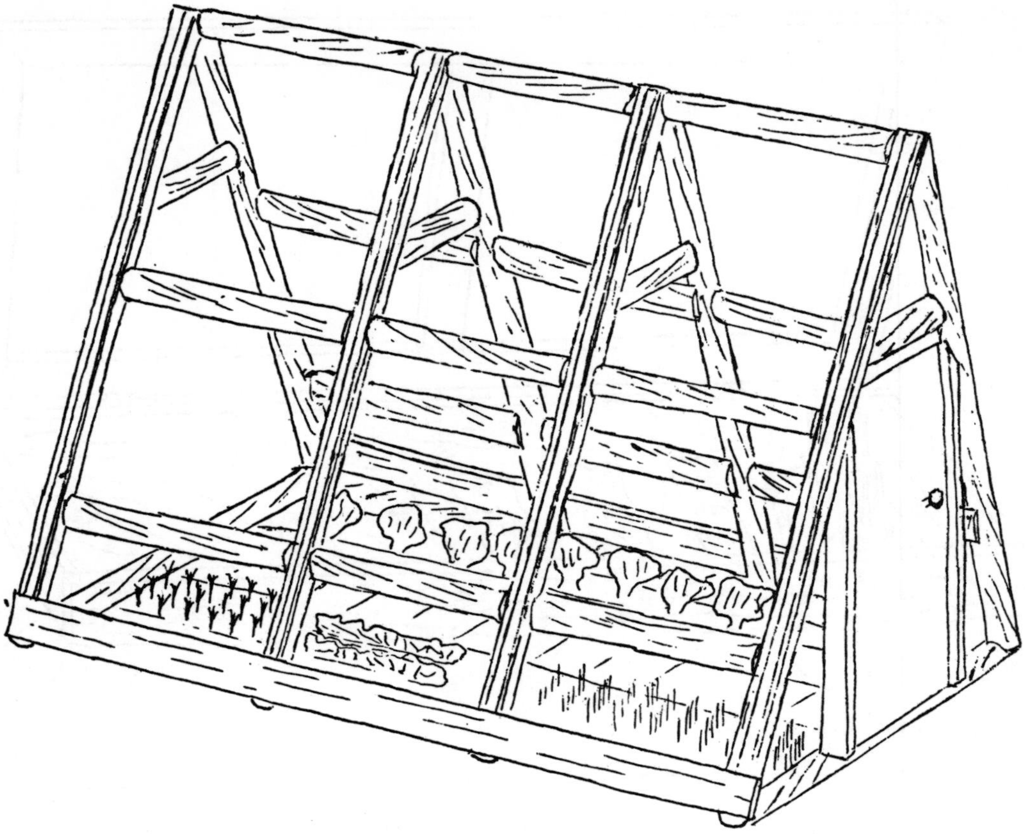

Fig. 2-7. A-Frame greenhouse.

two feet deep and one foot wide. Keep the topsoil and subsoil separated in two piles if possible. Stake ½-inch mesh chicken wire all around the outer wall of the trench to keep out rats and mice. Then fill the trench with horse manure, or whatever organic material is available. The reason for this is twofold. First, the heat generated by the decomposing organic matter will help heat the greenhouse and, second, it will insulate the sides and thus prevent frost from creeping in at the sides. When the trench is full, tamp it down well and use topsoil from the execavation for covering the manure to a depth of six inches. It should be heaped to compensate for settling. If no organic material is available to place in a trench, bury a heating cable all around the perimeter of the inside of the greenhouse to keep out frost.

Immediately inside the first trench dig a second trench seven inches deep and one foot wide. This is the plant-growing space and it is the reason for all the

rest of the structure. After the soil is removed, place one 48-foot length of soil-heating cable in each trench, making sure it is placed according to the manufacturer's directives. Next, cover the cable with about 1 inch of sand and then place ½-inch mesh chicken wire over the cable to keep the digging tools of the gardener from damaging it.

Finally, when all of this is in place, add compost or a potting soil mixture to finish filling the excavation. It also should be slightly mounded to compensate for the settling which is bound to occur. In addition, a supply of potting soil should be kept available for additional filling.

The center two feet of this greenhouse should be leveled very well and then covered with flat, thick rocks painted black. Suitable rocks can be picked up in fields, along roads or in creekbeds. Paint them with a good grade of black enamel. If dark stones are available, no painting may be necessary. The coloring is to provide a surface that will absorb as much solar heat as possible. Thus, the stones will hold the heat and radiate it during the night or whenever the sun isn't shining. During long periods of sunless weather the reverse can sometimes occur; then the rocks can be covered to prevent a chilling effect. Naturally, the stones also provide a walkway to keep the greenhouse operator from coming in contact with the soil and plants. As you gain experience in operating this greenhouse, you may want to eliminate the center walk and use it for growing space. This is fine, especially in the warmer climates. Just place boards in strategic locations around the floor.

When heating cable is used, the soil must be kept moist or it will act as an insulator for the cable and little good will be realized from it. Naturally, since we are growing plants in the soil directly above it, the soil must be kept moist for the plants also.

When snow comes, it can be banked outside the greenhouse to prevent frost from going down in the ground at the edges of the greenhouse.

Now, with floor and foundation all prepared, the frames can be put in place. It takes two people to set the frames up and nail the crosspieces to them. Each end of each frame should be placed on a flat rock or cement block to prevent it from sinking into the ground too far, which would slant the structure. A 1 × 12 board should be used all around the frame where the plastic contacts the ground.

Apply the fiberglass covering vertically over the frames. As it is being installed watch for protruding nails or wood splinters which could cause damage. Initially staple or tack the covering on the rafters. When it is all installed, nail 1 × 2 wooden strips over the seams on the rafters. Where possible, the covering should be doubled at the seams. Waste from the rolls can be utilized at the ends of the greenhouse.

Now, some supplementary heat should be

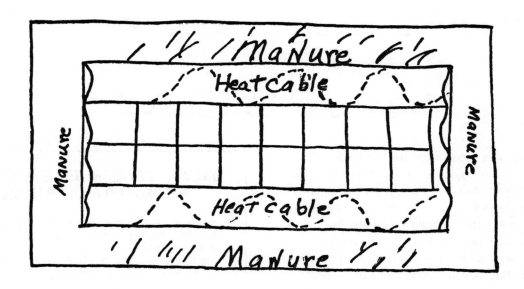

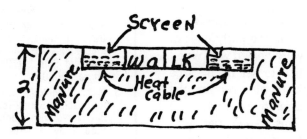

Fig. 2-8.   Below-ground layout of A-frame greenhouse interior.

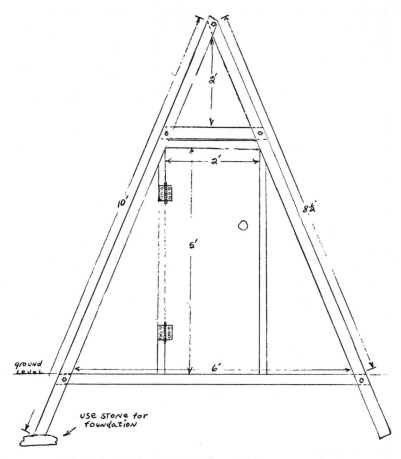

Fig. 2-9.   End frame with door in A-frame greenhouse.

available for this greenhouse for the coldest part of the winter. If the object is merely to prevent plants from freezing on the coldest nights a 5000 Btu. heater should do the job. It can be electric, gas, or oil. However, using the formula employed by heating contractors (see box), a 19,000 Btu. heater would have to be used to keep the plants growing steadily in the coldest climate at the coldest time of the year.

Heating a greenhouse is a fascinating subject. Over the hundreds of years that greenhouses have been in existence everything from hot springs to infrared heaters has been used. In the early days of American greenhouses many were heated with wood or coal. It was general practice to set up a space heater in the greenhouse with the pipe running the full length of the greenhouse to remove as much heat as possible from the smoke. This created problems with carbon monoxide and other gases circulating among the plants instead of fresh air. An improve-

ONE WAY TO FIGURE THE SIZE HEATER NEEDED FOR A GREENHOUSE

1. First find the temperature difference. This is the difference in degrees Fahrenheit between the lowest outside temperature and the temperature you want to maintain in the greenhouse. For instance, if you want to maintain an inside temperature of 60° and the coldest night temperature is −10°, the temperature difference is 70°.

2. Multiply the temperature difference by the number of square feet of exposed plastic or fiberglass. Include the roof, sides, and ends. Example: A 14 × 24-foot greenhouse has 744 square feet of area. Multiply the result of Step 1 (70) by 744. This will produce the figure 52,080.

3. If the greenhouse is covered with one layer, multiply the result of Step 2 (52,080) by 1.2. If the greenhouse is covered with a double layer, multiply the result of Step 2 by .8. Read the answer directly in Btu's. Example: 52,080 × .8 = 41,664.0 Btu's.

ment was to install the space heater outside the greenhouse in a shed of its own and run the smoke pipe the length of the greenhouse to a chimney located on the opposite end.

Wood and coal fires require frequent tending which sometimes involves getting up in the middle of the night, and there is the ever present danger that the fire will go out on the coldest night and let all the plants freeze. Many people have used wood or coal-heated greenhouses for years without a single accident, however, and a person operating on a limited budget could use a large wood stove; even make one from an old oil drum and burn waste wood, slabs, roots, coal, or whatever was available to heat his greenhouse at almost no cost. Accidents could be minimized by installing a warning device that would ring an alarm inside the house if the temperature fell too low. These devices will work with batteries if no AC electricity is available.

In our small A-frame greenhouse all sorts of combinations are possible. A wood stove could be used to heat it during very cold days and a portable kerosene heater set up to heat it at night in addition to the soil-heating cable. More expensive alternatives would be thermostatically controlled gas or electric heaters. In all cases some warning device should be used to signal dangerously low temperatures.

Equally important to heating the A-frame greenhouse is ventilating it. Since it is small and the door at one end proportionately large all that is generally necessary is to leave the door open for a few minutes each day, probably while the plants are being tended. A good indication of the condition of the air in the greenhouse is the odor. If it begins to smell musty, ventilation is needed. On spring or fall days some shading should be provided to prevent the plants from sunburning. A very good shading material to use with the A-frame greenhouse is cheesecloth

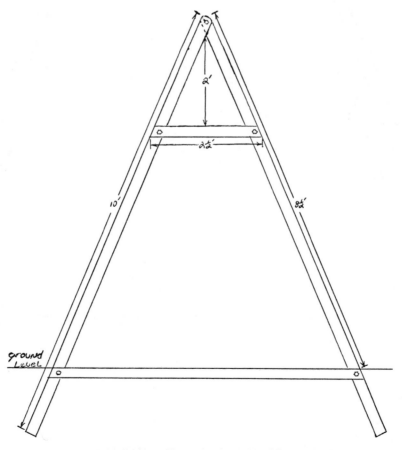

Fig. 2-10. End frame without door in A-frame greenhouse.

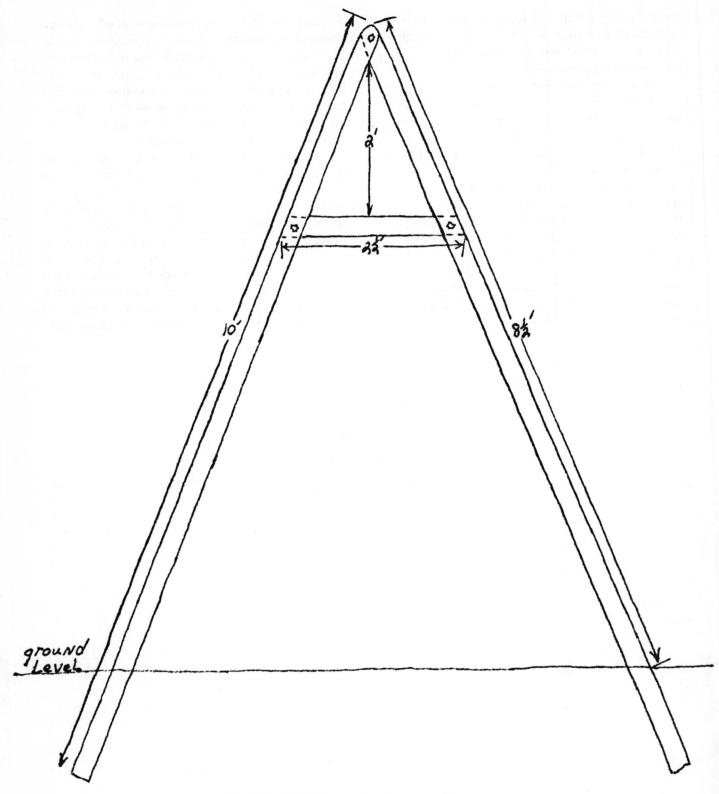

Fig. 2-11.   Middle frame in A-frame greenhouse.

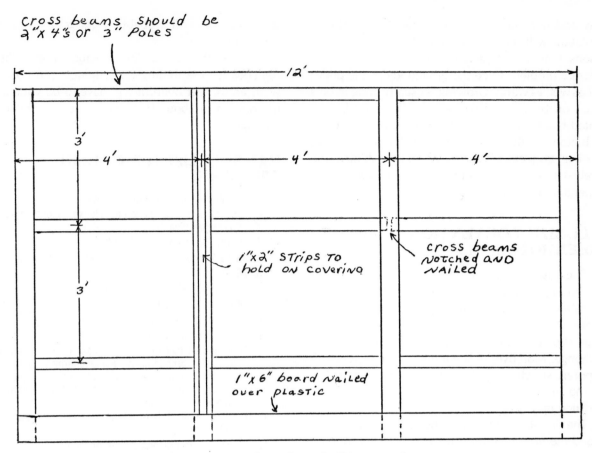

Cross beams should be
2"x 4's or 3" poles

3'

4'

4'

4'

3'

1"x2" strips to
hold on covering

cross beams
notched and
nailed

1"x 6" board nailed
over plastic

Fig. 2-12.   Structural members of A-frame greenhouse.

stretched over stakes driven in the ground at each side of the plant rows inside the greenhouse. Cheesecloth allows enough light to promote plant growth while at the same time preventing an excess of heat and light which can cause sunburning.

Lest this discussion makes it seem as though it takes a great amount of time to tend this greenhouse let me point out that we average about five minutes a day in ours. Many times it only has to be looked after once a week, when it is watered.

Watering greenhouse vegetables should be done very carefully, especially in winter, so no moisture splashes on the leaves. This can cause mildew. It is better to use a perforated hose extending from a watering can to slowly soak the soil at the ground level.

EXCAVATED GREENHOUSE

The excavated greenhouse needs little or no supplementary heat since it uses the heat from the ground as well as the heat from the sun to keep the frost away. The first step in building one is to excavate a basementlike structure of whatever size desired for the greenhouse. Drainage is of primary importance here or else you might wind up with a fishpond instead of a greenhouse since the excavation should be about six feet deep. Unless the side walls of the structure are very dry and firm, they should be shored up with cement blocks or treated timbers to keep the walls from caving in and to keep moisture out.

Sometimes old basements of houses that have burned down or been otherwise removed are good places to start a greenhouse of this type. Also on some farms and country places caves exist that face the south and are warm the year around. They, of course, are almost ideal.

Regardless of the excavation used, the top or entrance should be covered with clear plastic or glass to keep the cold out and the sunshine in. Vertical excavations should be covered with a peaked roof to shed snow and rain. Generally, some means of artificial ventilation will have to be provided in the warmer seasons with this type greenhouse but cooling is seldom a problem since the ground keeps the plants cool as well as warm. Growing is usually done in pots placed on benches. If space is available oil barrels painted black and filled with water can be placed in the bottom of the excavated greenhouse to act as solar heat collectors. When the sun isn't shining, the barrels will radiate heat and help to maintain an even temperature inside the greenhouse.

FULL-SIZE CONVENTIONAL FRAME GREENHOUSE

The last project of this chapter is a modern greenhouse, twenty-four feet long and large enough for raising a considerable number of plants, either for private use or a small commercial venture. The design should pass most building codes with very little change. Moreover, since it might be considered a temporary structure by the tax assessor, it is possible that no increase in taxes would result from its construction.

The first step is to find out how far from the street or road the front wall must be located. Your friendly building inspector is the person to consult for this information. Generally, it will be the line of buildings already constructed. When this is known, the foundation site can be located and measured in.

Lay out the front line of the foundation by measuring equal distances back from the street or sidewalk in two locations twenty feet apart. Drive stakes at these points and stretch a chalk line between the stakes. Next, stand between the chalk line and the street, facing the chalk line, and determine where you want the left side of the building to be located. Mark this on the chalk line by folding a small piece of tape around the line. Use a plumb bob to find the point directly under the tape. Drive another stake there, attach a chalk line to it, and stretch it at right angles to the first line to locate the lefthand wall. Use a carpenter's square to make this corner as square as possible. Now continue to use the plumb bob, square, and additional chalk lines to lay out the entire foundation. Finally, square it up by measuring diagonally from one corner to the other and adjusting the rear wall line until the diagonal measurements are

the same. You will probably never get them perfect; no one ever does.

When the foundation is squared up, dig the trenches for the footings. The trench can be 1 foot wide and 4 feet deep. It will have to be widened at the bottom to about 16 inches to accommodate the footings.

MATERIALS LIST FOR FULL-SIZE GREENHOUSE

1. 26 cubic feet concrete for footings
2. 675 12 × 6 × 8 concrete blocks. Lightweight is recommended.
3. 70 running feet of 2 × 6 for sill
4. 38 43.5-inch 2 × 4's for studding. 48-foot studs.
5. 76 feet of 2 × 4 for wall plate
6. 24 running feet of 2 × 6 for ridgeboard
7. 26 8-foot 2 × 4's for rafters
8. 12 feet of 2 × 4 for end frames
9. 2 doors, frames and hardware.
10. 1600 square feet of .037 thickness flexible fiberglass made for greenhouses. For economy polyethylene can be used and replaced gradually by fiberglass to spread costs over a long period. Rigid fiberglass panels are also available.
11. Heaters, exhaust fans, shutters, electric wiring, and water pipe
12. 2 cubic yards crushed gravel for the floor. 48 running feet of 2 × 10 planking for walkways.

If the soil is firm no form will have to be built for the footings since the soil can be shaped properly to receive the concrete. A good concrete mix for footings of this type is 1 part portland cement, 2½ parts sand, and 3½ parts gravel. The footings should be 3 inches thick. The concrete for this footing can be mixed in a wheelbarrow or mortar box.

When the footings are curing, which will take three days, you can be installing the water and electric lines. Water lines will have to be buried at least 4 feet or protected by heating cable. Water lines could be ⅝ or ¾-inch diameter plastic or copper. The electric line should be #2, 3-wire cable insulated for burying. Connections at either end might have to be done by a licensed tradesman.

Both water and electric lines can be buried in the same trench if expedient. This digging also can be done by hand. Dig under the footings and don't forget to leave the ends of the lines long enough to make the connections inside the walls.

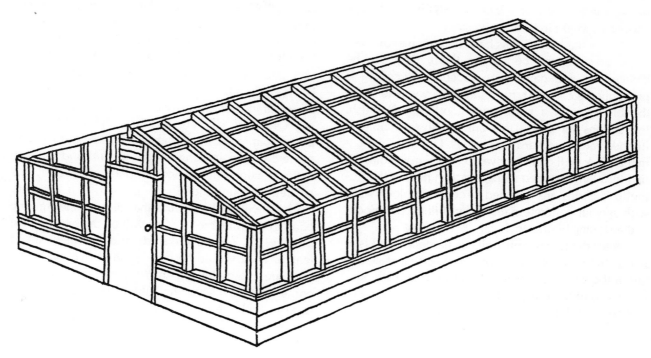

Fig. 2-13. Full-size conventional frame greenhouse.

After the footings have cured properly, the concrete block wall can be started. Lightweight blocks should be used if they are available, since they have greater insulating value. Bags of mortar mix are available for block laying, and generally it is easier and cheaper to use this mix than to mix your own. Besides the blocks and mortar you will need a trowel, carpenter's level, chalk line, and a pair of gloves.

Start by laying the corner blocks. Notice that the corner blocks have one flat face. This, of course, is laid to the outside. Set the first block on a full width of mortar about an inch thick. Level it up and square it with the footings very well since this first block determines the line for the rest. Do this at each corner. By the time you finish, the mortar should have stiffened enough to permit placing the chalk line between the first and second corner blocks. Use this as a guide to place three additional blocks in each direction. Notice that the blocks overlap to tie the corners together. In addition, it is well to fill the holes in the corner blocks with mortar and push a concrete reinforcing rod from top to bottom of the wall after the blocks are all in place to further reinforce them.

All blocks except the corner blocks should be filled with insulation.

Lay mortar bead 1 inch thick the width of the face shell of the block. In addition, the inside end of the block should be "buttered" with a similar bead of mortar. Push the blocks down and level them out so the joint is about $3/8$ of an inch thick. Be sure to level the block in both directions and keep the joint the same thickness. It is customary to build up each corner until only a corner block is left to be placed. Then the spaces between the corners are filled in. The door frame should be placed in position when the ground level is reached since it will be recessed three tiers of blocks. Half blocks are available for filling in sections where full blocks cannot be used. Also blocks can be cut to fit by scoring them with a chisel, or a masonry blade which will cut blocks can be purchased for your electric saw.

The cement block wall of our greenhouse extends two feet (three courses) above the ground level. Anchor bolts are placed in the top row of blocks for use in tying the wooden frame to the cement blocks. The wall then should be allowed to cure before further work is done on it. Curing will take about two weeks in summer weather.

When the block joints have hardened enough so some pressure can be put on them, they can be cleaned up and painted or stuccoed. Stuccoing is probably the cheapest, most satisfactory way to cover the blocks with a permanent good-looking finish. Instructions for doing this are readily available at lumber yards and hardware stores. Likewise, the cement block wall can be covered with wood siding by first installing wooden strips and then nailing the siding to it. Strips are attached to concrete block walls with special concrete nails or by drilling the blocks and installing plastic anchors.

After the foundation wall has hardened and is in place, the framing can be put up. The first step is to fasten the sills to the wall. The sills should be a double 2 × 6 of knot-free softwood such as redwood or treated fir. Bolt the sills to the wall by drilling holes to correspond to the location of the anchor bolts placed in the wall. At the corners the plate should be cut to overlap so the opening in the joint doesn't extend from inside to outside. The sill should be treated with two coats of a good preservative and the joints between the sill and the wall should be caulked with a good grade non-hardening caulking compound.

Once the sills are in place, the studding and other framing can commence. Professional carpenters do the framing on the ground and raise an entire section into place at once. However, the amateur working alone with perhaps little space to work in will probably find that placing the framing boards one at a time in place will be the easiest. First procure 38 2 × 4's for studs and cut them all to length. They should be cut 43.5 inches long. When they are cut to length, measure off and place the corner studs in place. Plumb these corner studs with a level in both directions and brace them in place by

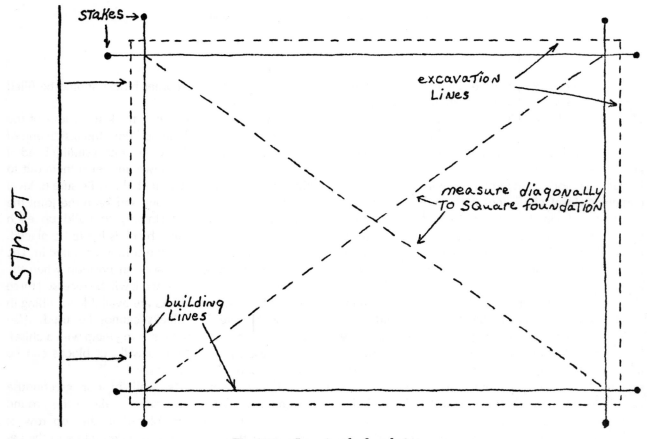

Fig. 2-14. Squaring the foundation.

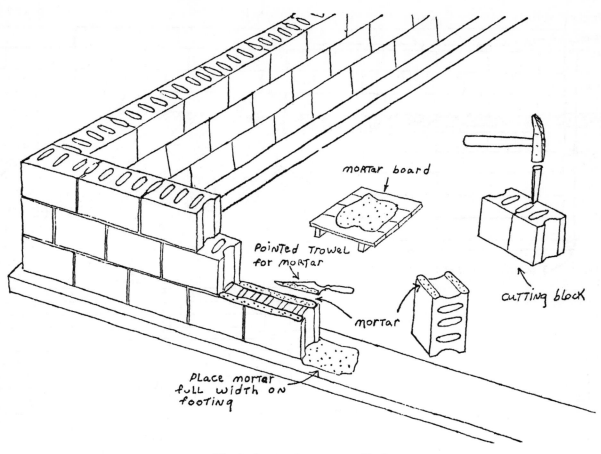

Fig. 2-15. Laying concrete blocks.

tacking a 1 × 4 from the stud to the sill. Next stretch a chalk line 3 feet off the sill from the outside of one corner stud to the other all around the building. Next find the center stud, mark on both sides of the building, and install a double stud by nailing two studs together. This is necessary because of the top plate being spliced at this location. Plumb these center studs also and brace them in place.

Now, the top plate can be a 12-foot 2 × 4. It will of course, take two to each side. Nail them in place with the center joint evenly spaced on the center stud. All that remains to finish framing the sidewalls is to nail the other studs in place 2 feet apart. The top plate and sill will indicate the correct location.

With the side wall studding in place the end wall studs can be considered next. Note that the two center studs which will be on each side of the door frame are 8 inches longer than the other studs. Toenail these in place and then install the top plate,

which is cut 66 inches long. Across the top of the doorway nail a 3-foot-long 2 × 4 to complete the frame. Both ends are the same.

The next step is to install the roof rafters. First nail the ridgeboard in place. This should be a 2 × 6, 24 feet long if available. If it isn't, use some combination of lengths that will avoid having the joint in the center such as would be caused by having two 12-foot lengths joined at the center.

Temporarily nail the ridgeboard in place, directly in the center of the building. Use a short length of scrap 2 × 4 to hold the ridgeboard 48 inches above the door frame on each end. The rafter detail is shown in Figures 2-13 and 2-18. It is important that the rafter heel cut fit the plate exactly so it won't extend past the side rafters, which could cause damage to the covering. Lay the rafters on 24-inch centers and toenail in place, using 16D nails on each side of each rafter.

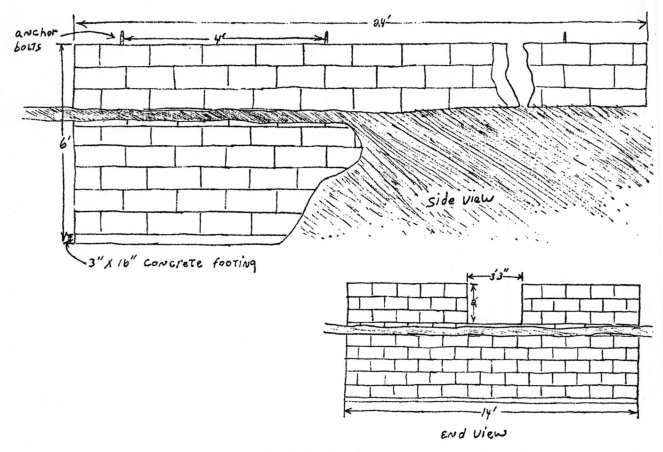

Fig. 2-16.   Foundation.

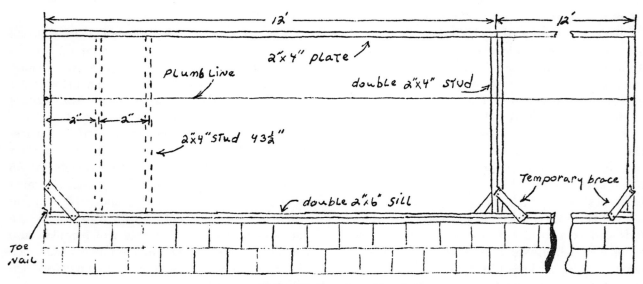

Fig. 2-17.   Installing the wall studs.

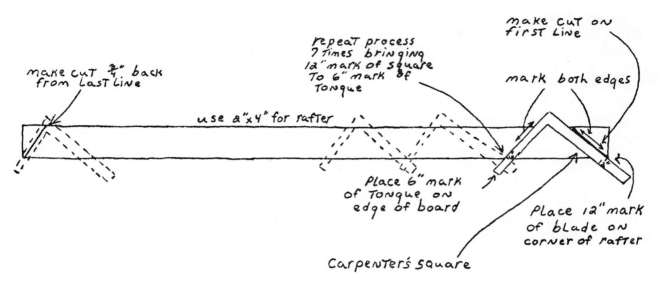

make cut ¾" back from last line

repeat process 7 times bringing 12" mark of square to 6" mark of tongue

make cut on first line

mark both edges

use a"x4" for rafter

Place 6" mark of tongue on edge of board

Place 12" mark of blade on corner of rafter

Carpenter's square

Fig. 2-18.   Master rafter.

Forming the angles for the rafters can be trying for the novice carpenter. Generally it is best to make a master rafter and use it for marking the rest. This greenhouse rafter pitch is called a ¼ pitch since the rise in the center is ¼ of the width of the building. Further, by comparing the span by the length of the rise we find that the roof rises 6 inches for each foot of span. This is all the information we need to find the correct pitch for the rafters. First select an 8-foot 2 × 4 to use as a master rafter. Square one end by sawing if necessary and place it on two sawhorses where it will be comfortable to work with. Now take a carpenter's square and hold it so the blade (2-foot part) is in your right hand and the tongue in your left hand. Apply the square to the rafter so the 12-inch mark on the back surface of the blade coincides with the corner of the 2 × 4 nearest you. Further, pivot the square so the 6-inch mark on the tongue lines up with the edge of the rafter nearest you. Use a hard pencil to mark across the 2 × 4 along the back edges of both the tongue and blade. Now move the square to the left and place the 12-inch mark of the blade on the pencil mark where the tongue 6-inch mark was previously located (see Fig. 2-18). Also, as before, line the 6 inch mark of the tongue with the edge of the rafter. Notice that this places the square in exactly the same position as before except it will be moved over about 13⁷/₁₆ inches. Repeat this procedure seven times, marking each line each time. The first pencil mark you made will be the line for the wall plate and

the last will be the line for the ridgeboard. Now, very carefully measure and mark a line ¾ inch back from the previous ridgeboard mark. This will be the final ridgeboard mark. Finally saw the 2 × 4 on the final ridgeboard mark and on the opposite end at the wall plate mark.

This rafter can be used for marking all the rest, but be sure to put it in position before you mark any more to see if it is correct. After the roof rafters are in place install the 2 × 2 purlins and side rails. The purlins are made from 2 × 2 stock to minimize shading; however, they can be made from 2 × 4's too at the discretion of the builder. At each end over the doors a 21-inch-square opening for the exhaust fan and shutters is provided. Note details in Figure 2-19.

When the framing is complete the fiberglass covering can be applied. Notice that it is applied horizontally on the sides of the greenhouse and vertically on the roof. The waste from the roof can be utilized at the ends. Make sure the covering is pulled tight to eliminate wind rippling. Greenhouse fiberglass usually comes in 4-foot rolls; use 1 × 2 wooden strips or the equivalent to fasten the edges of the rolls to the rafters. Whatever waste is left from the fiberglass covering the greenhouse can often by utilized in building cold frames. The door can be solid or a door with windows. In the winter a considerable heat savings can be realized by building an entryway around the entrance doors.

The inside of the greenhouse must be covered

with polyfilm plastic to create a double wall in cold climates. This will effect up to a 40 percent heat savings. The polyfilm of course, is fastened to the inside of the 2 × 4 rafters. Use only the polyfilm made especially for greenhouses.

When this is done, the exhaust fan and shutter can be installed in the opening previously framed in. In addition, the heater or heaters should be installed. This size greenhouse will need a 60,000 Btu. heater to maintain the 60-degree air temperature in −20-degree weather. Actually, most operators opt for a heater much smaller than this and content themselves with providing supplemental heat just to keep the plants from freezing during the coldest days, since as soon as the weather moderates the greenhouse will warm and the plants start growing again. However, both the exhaust fan and heaters should be equipped with an automatic temperature regulating thermometer. In addition, every greenhouse should have

an automatic alarm system to warn the operator of too low or too high temperatures. Automatic misters and waterers can also be installed at the discretion of the operator after he gains experience. Level the floor and cover it with 2 inches of pea gravel. Walks can be formed with concrete blocks or wooden planks.

When the greenhouse construction is finished, the "furniture" can be installed. This model greenhouse can use a 3-foot growing bench on each side and a 4-foot bench in the center. The space under the benches can be used for growing plants with supplemental lighting or it can be used for storing potting soil, tools, and other necessities of greenhouse operating. Don't forget that hanging pots can be used also to raise plants if more space is needed.

No attempt will be made to give instructions for setting up the plants in the greenhouse, since this will vary considerably among operators. Notice that this

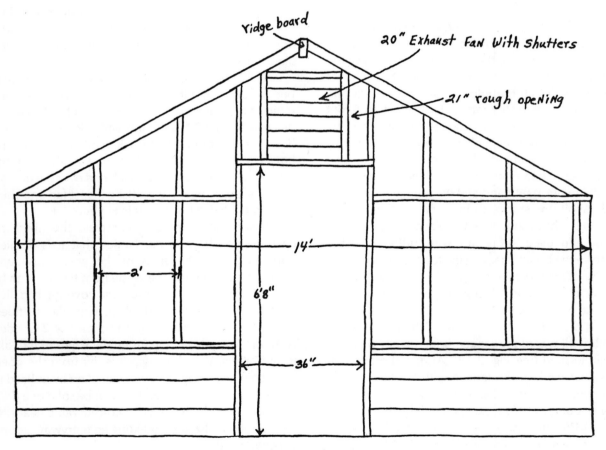

Fig. 2-19.   End view.

greenhouse can be made longer with no loss of symmetry if conditions warrant it at some later date. Also, it is highly adaptable to whatever personal innovations you may care to incorporate.

Control diseases and insects by keeping the doors shut tightly and sterilizing all tools periodically. Also, try not to introduce any new plants to the greenhouse without a two-week quarantine period and a careful inspection.

Some tools that are useful in the greenhouse are the dibble board and dibble. A dibble board is a board with evenly spaced holes drilled in it. The dibble is a pointed stick. They are used by laying the board on a seed flat and poking evenly spaced holes with the stick in the seed flat to plant seeds in. A masonry trowel for transferring soil and fertilizer is also useful as is a tamping block, rubber bulb for watering plants, hand sprayer, hand cultivator, scoop, sieve, and sharp knife. All of these tools can be handmade if you don't desire to purchase them.

If none of these greenhouse structures exactly fit your ideal, use the ideas mentioned to design your own. Also, almost every need of the prospective builder can be satisfied from commercial sources these days. Following is a list of greenhouse suppliers. No specific recommendation of the companies listed below is given or implied. The addresses are listed only as a convenience to the reader. See the classified ads in gardening magazines for other company names.

Environmental Dynamics
P.O. Box 996
Sunnymead, California 92388

Peter Reimuller
Post Office Box 2666
Santa Cruz, California 95063

Turner Greenhouses
Post Office Box 1260
Goldsboro, North Carolina 27530

McGregor Greenhouses
1195 Thompson Avenue
Santa Cruz, California 95063

Greenhouse Specialties Co.
9849 Kimker Lane
St. Louis, Missouri 63127

Redwood Domes
Aptos, California 95003

Sunshine Greenhouses
P.O. Box 3577
Torrance, California 90510

Greenhouse gardening is a fascinating but challenging occupation. No one ever learns enough about it to produce guaranteed results but it is always possible to learn more by reading the many excellent publications dealing with the construction and operation of greenhouses available today. The U.S. government, in particular, offers a wealth of valuable information. The following pamphlets should prove helpful:

Building Hobby Greenhouses
Agriculture Information Bulletin No. 357
U.S. Government Printing Office
Washington, D.C. 20402

Sash Greenhouses
Leaflet No. 124 U.S. Department Of Agriculture
Superintendent Of Documents
U.S. Government Printing Office
Washington 25, D.C. 20402

Also write to *United States Department Of Agriculture, Agricultural Research Service, Crops Research Division*, Beltsville, Maryland 20705 and ask for the publication *List of Sources of Information on Greenhouses*. This is twelve pages of sources of every type of greenhouse information.

The many challenges of greenhouse operation keep its practitioners on their toes. Maybe that's why there are so few ex-greenhouse operators.

Chapter 3

A THREE-BEDROOM A-FRAME HOUSE

We have spent several happy years living in an A-frame that we designed and built ourselves and we have noticed several outstanding qualities. Possibly one of the most endearing qualities of an A-frame is you don't have to be an expert carpenter to build one. Also, the square footage of lumber required to enclose the space is less than on any other type of structure except a geodesic dome or some other very radical designs which have some not so endearing qualities. Also, an A-frame lends itself so well to using huge sturdy beams, lots of stone, and rough irregular natural shingles, siding, and other rustic materials that it is an innovator's delight.

One other feature which makes an A-frame practical is that it can be set on piling; it can therefore be used in rough country without leveling the ground. A hillside is a fine location. Set an A-frame up among the trees and it blends in so well it almost becomes a part of the forest. Live in one of these A-frames in the trees and you feel almost like a forest creature yourself.

The A-frame design, of course, is not new. It is reportedly a gradual assimilation of designs used in Scandinavian countries. The steep roof refuses to hold up snow and our roof releases layers of snow with a great "whoosh" whenever it gets more than two inches. This snow slides down the roof and piles up around it underneath, thus creating a bank to prevent the winter air from blowing underneath the floor. In fact, if the roof did not pile snow there naturally, we would have to get our old shovel and pile it in there artificially.

The inside also lends itself to finishing with economical materials, and the need for skilled plasterers and carpenters to finish the inside is eliminated.

One objection to an A-frame we hear frequently is that the walls slant in. True, they do slant and that's the way it should be. After living in one awhile you begin to wonder why builders ever built walls straight up in the air in the first place. The landscape painted by your favorite aunt will in fact be placed on a better slant for viewing from a sitting position in an A-frame than it would be in a conventional straight-walled building.

Realtors and insurance men also seem to take delight in an A-frame and no trouble at all will be experienced in selling or insuring a well-made A-frame placed where such a building should go.

Now, that is the bright side of building with such a design. The negative side is omnipresent also.

One disadvantage of A-frames is that hardly anything is custom-made for them. Many suppliers do not have doors narrow enough to look good with the lines of an A-frame; many do not have the proper size windows. Also such things as hot and cold water tanks, electrical boxes, conventional furniture, electrical appliances, and kitchen cupboards were never designed with an A-frame in mind. However, all things can be adapted to use with the A-frame by applying a little free thinking. So let's start building.

SITE SELECTION

First select the site. In the author's opinion it would be a pity not to build an A-frame facing directly east and west. To see the sun rise and set is one of the greatest joys of life and especially so from the top story of an A-frame. A good case could be made for facing the A-frame north and south in areas of heavy westerly winds since storms originating from directly east or west would then expend their fury on the well-shielded roof instead of against the ends. However, with a north-and-south orientation the sun is apt to glare in the south windows during the hottest part of the day and the north end is likely to receive a very low intensity of light, especially in winter. Of course, an artist would probably prefer the northern light.

DRILLING A WELL

In addition to light, water is an important consideration. If you are going to live full time in an A-frame, you certainly should take thought even before you plow away a single rock or cut a single tree, or for that matter even buy the lot, about where you are going to get water. In other words, drill the well first or have some other way of getting water.

Many beautiful houses were completely built before the well was drilled but when it came time to drill the well no potable water could be found. Equally tragic is the case of a homesteader I know of only a few miles from our place in northern Wisconsin (where water is taken for granted) that built his cabin first and then drilled the well. To date, he is on his third hole with a huge well-drilling figure to pay; his water is "bacteriologically unsafe."

However, this doesn't mean that you should drill a well just any place on your homesite. Make sure that you know where the house is going to set and drill the well if possible so it doesn't require a long trench to bring the water into the house. Generally speaking, the depth of a well determines how safe and how reliable it is. As a rule, the deeper the better, although, like all rules, this one has exceptions. Also, depending upon the depth and soil composition, it is frequently possible to bring in your own well, especially if there are no regulations against it. It might be well to check into this before you drill. Also, try to make sure that the water line won't be placed under a frequently used path. In cold climates just walking across the top of the dirt where the water line is placed can drive the frost down to the point where the line will freeze up. Frost under traveled paths sometimes goes as deep as ten feet, whereas frost in an unpacked area only goes to a foot or so.

If you drill your own well, make sure it is located where it won't be contaminated by an outdoor privy, stockyard, or barn or even your own septic system. Generally, this means the well should be located on ground higher than the source of possible pollution.

SEWAGE DISPOSAL

Disposal of sewage also deserves consideration, especially in rural areas. In many states a rigid code regulating the placement of septic tanks has been adopted. In such states they have to be put in by a licensed plumber who is designated by the state as a septic tank installer. Septic tanks can't be installed too close to lakes or other water sources. So it is well to give some thought to this when you select your building site. One method a builder who desired to place his own septic system could use is the holding tank. A holding tank is a large sealed concrete or metal container that stores the effluent from the house sewer system without dispersing it throughout the ground as a septic tank system does. Periodically, it must be pumped out. Contractors with huge tank trucks perform this service in some areas, usually visiting the holding tanks three times a year. The cost is generally low and this method is well worth considering if you want to build in proximity to a lake, creek, or river.

Also, don't forget the type of bathroom stool that decomposes all the effluent right in the bathroom without ever using water. This makes it possible to get along without a septic tank; in such a case all you need is drainage from the sinks and baths to a dry well or seepage system underground. Many regulations permit this. Hardy folks also can give consideration to the outdoor privy, which has fallen into disfavor primarily because of some easily correctable faults. Some of these are odor, distance from the house, and the "chill" factor. See my book *The Homesteader's Handbook* for ways to correct this.

ELECTRICITY

Another important point when you are deciding on your homesite is the availability of electric lines. Make sure before you build beyond the high line that you don't mind getting along without electricity. If you must use a home generator with a gasoline engine, it will probably cost more and cause more pollution than using the public utility. Most power and light companies will charge around $2000 a mile to bring a power line to your place and in addition they must be guaranteed a minimum monthly bill for five years. However, wind-operated home electrical systems are being developed rapidly. Possibly in the near future most anyone can have his own electrical system.

ACCESS TO ROADS

Also check very closely your access to a public highway. Many times, especially in remote regions, the roads have been built over private land according to a very informal agreement, possibly even just a "handshake." While the principals still own the land the agreement may be good. After they sell or die, the next person may not agree to the terms, thus sealing your land off from a public roadway. Also, it is not wise to assume that your attorney or mortgage company will take care of this. They might not. Remember it is the buyer who will be liable for all charges whether or not a law suit develops over access.

With all the annoying but important details taken care of, the fun of building can commence. The first step is to cut the trees and brush and get rid of any large projecting rocks on the building site. This building sets eighteen inches off the ground at its lowest point; so there is no need to level the ground underneath. However, make sure there is a slope; you don't want rain puddles standing for weeks under the building. You especially don't want them standing around the footings since they could weaken the foundation. Also take time to dig two or three holes in your proposed building site to make sure the subsoil will support a building.

LAYING OUT THE BUILDING

The first step in laying out the building is to drive a stake at the proposed lefthand front corner. Then use a magnetic compass to extend the outline for the lefthand side of the building directly west of this point. Measure off the entire outline, which is 30 feet wide and 44 feet long. Use a chalk line and stakes to designate the outline.

When all is done, use the outline to determine if this is truly the direction and the location that you want for your new home. If it is, proceed to lay out the location for the footings on which the building will sit. The footings will be made of concrete and the frames will be anchored to each. It is expected that the building will have no basement.

FOOTINGS

This building rests on twenty-four concrete footings, twelve on each side. In laying out the footings care must be taken that they are aligned both lengthwise and crosswise. Also, extreme care must be taken that the excavations for the footings are made in the true locations.

Find the true locations for the marking stakes by stretching a taut chalk line along the south side. Carefully place the first stake and the last stake. The stakes should be 44 feet apart. Then measure the location of each intermediate stake, which will be at 4-foot intervals. When this is done, the problem of how to find an exact duplicate of this line "squared" up exactly with it but separated from it by 30 feet takes precedence. We solved this problem by fabricating a huge "square" formed of two 8-foot 2 × 4's. Then we propped the "square" up so it was perpendicular with the already established lefthand line and with the help of a level, a chalk line, and a steel tape found the location for the righthand rear corner stake. When this was established we went to the opposite end of the building and found the exact location for the righthand front footing. With these two points established, we easily found the correct location for the intermediate footing by measuring the correct 4-foot distance.

After the correct locations were found we still had to project the desired elevation or level for each footing. Since we had previously decided to use 6 inches from the soil as a minimum elevation we simply used the highest point at which a stake was located as a grade point and measured up 6 inches on that stake. We used that point as a grade mark and used a line level and chalk line to project all of the other elevations on all of the stakes. Our location only varied about 6 inches we found, after all the marks were designated on all the stakes.

Now the unevenness in the ground level can be overcome by two methods: either the frames of the building can be made longer or the footings can be made higher. We elected to adjust the size of the footings since we decided sometime in the future we might landscape the ground around the building to a level point. The footings are 16 inches square and they are set in the ground at least 24 inches, or to subsoil. The face of the footing that has the bolts must be slanted to the incline of the frames (see Fig. 3-2). Two ½-inch machine bolts are heated and bent and placed in the concrete when it is wet. They are set in the center of the incline of the footing so there is a 6-inch space between them. It is recommended that the bolts be set 3 inches into the concrete.

Also at this time additional hardware might as well be made up. An angle bracket which bolts to the frame and to the bolts placed in the concrete is used. This angle bracket should be made from ¼-inch strap iron drilled with 9/16-inch holes at the proper locations. All of this work can be done at home or it can be done by a blacksmith.

MATERIALS LIST FOR A-FRAME HOUSE

Foundation
1. Approximately 64 cubic feet concrete for footings
2. 48 ½ × 6-inch hex head machine bolts, nuts, and flat washers
3. 48 ½ × 3-inch strap iron angle brackets

First/Second-Floor Framing
4. 96 ½ × 6-inch carriage bolts, nuts, and washers
5. 20 pounds thirty-penny nails
6. 48 14-foot 2 × 6's for first-floor rafters
7. 48 12-foot 2 × 6's for second-floor rafters
8. 12 10-foot 2 × 6's for filler pieces (first floor filler 24-inch, second floor filler 30-inch)
9. 12 14-foot 2 × 12's for floor joists
10. 20 12-foot 2 × 12's for floor joists
11. 12 1-foot pieces for closing peak of second-floor rafters (Use scrap if available.)
12. 12 2-foot 2 × 4's for bracing upstairs rafters
13. 143 8-foot 2 × 4's for cross framing between rafters
14. 120 4-foot 2 × 6's for first-floor cross joists
15. 70 4-foot 2 × 6's for second-floor cross joists

Sheathing and Flooring
16. 66 4 × 8-foot sheets ⅝-inch exterior plywood for roof sheathing
17. 38 sheets ⅝-inch exterior plywood for first- and second-floor subfloors
18. 38 sheets ⅝- or ¾-inch particle board for underlayment
19. 15 pounds eight-penny coated nails
20. 10 pounds threaded flooring nails (eight-penny coated can be used)

Framing for Ends
21. 22 8-foot 2 − 6's for end framing

End Sheathing
22. 10 4 × 8 sheets ⅝-inch exterior plywood for end sheathing

Downstairs Porches
23. 48 8-foot 2 × 6's
24. 14 8-foot 4 × 4's
25. 2 12-foot 1 × 12's for steps

Upstairs Porch
26. 7 4-foot 2 × 6's
27. 2 4 × 8 sheets ⅝-inch particle board or exterior plywood

The concrete mix should be the traditional 1-2-3 mixture composed of 1 part concrete, 2¼ parts sand, and 3 parts gravel. In addition, reinforcing steel such as wire or scrap iron can be incorporated into the mix.

If for some reason it is not desirable or possible to use concrete footings, posts can be utilized. They should be treated cedar posts, about 10 inches in diameter, and they should be sunk to a very firm subsoil or even to rock if possible. If the soil is sandy or perpetually wet, a footing of rocks should be provided for the post. This can be set in concrete also if desired. If the ground is too wet, you have no doubt selected the wrong location. Better look for a different site.

After the footings are placed, they should be allowed to set 3 weeks before any weight is put upon them. Use this time to build the frames.

FRAMES

The frames are built from 2 × 6 lumber. These are some of the most important structures in the building so be sure you use only well-seasoned pine or fir. Also, they should be knot-free. If possible, use only heartwood material. You will need 48 14-foot 2 × 6's for the rafters and 12 10-foot 2 × 6's for the filler pieces between the rafters at the second floor level and at the bottom. Additionally you need 12 14-foot 2 × 12's and 20 12-foot 2 × 12's for use as floor joists. They must be well seasoned and knot-free. Once you get the lumber home, lay it out so the individual sizes are in separate piles. Next, saw the 10-foot 2 × 6's into an equal number of 2- and 2.5-foot pieces. Separate these into piles also. Now you will also need the hardware. This consists of 96 ½ × 6-inch carriage bolts with nuts and washers, and 10

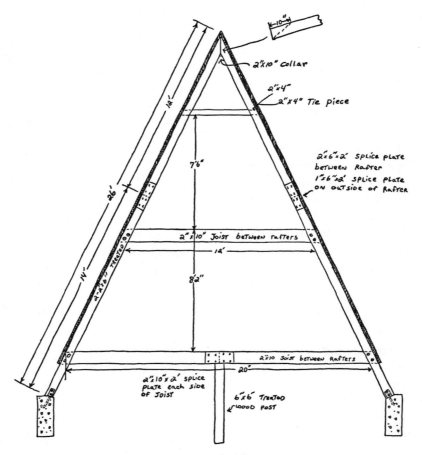

Fig. 3-1.   Typical A-frame.

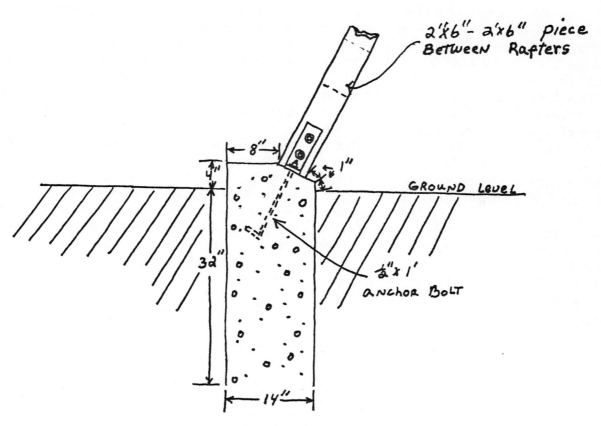

Fig. 3-2. Footing and rafter connection.

pounds of thirty-penny nails. For tools you need a hammer, brace and 9/16-inch wood bits, saw, and steel measuring tape.

First measure off and square the ends of two 14-foot 2 × 6's and lay them together. Then select one 2.5-foot filler piece and nail it in position flush with one end between the 2 × 6's. This will be the bottom of the brace. Next nail a 2-foot filler piece so it projects 1 foot from the other end between the 2 × 6's (see Fig. 3-2). Nail all of these in place with two thirty-penny nails in each end. Make sure that the fillers are flush with the surfaces of the rafters and also make sure the filler piece projection at the top end of the rafter is properly aligned with the horizontal and vertical centerlines of the rafters. If you have confidence in your carpentering ability, you can drill the holes at this time for the bolts which hold the floor joists to the rafters. Place the first bolt at the centerline of the rafter exactly 29³/₈ inches from the bottom end.

Drill the next hole 32³/₈ inches from the bottom and 1 inch from the inside edge. This completes the holes for the bottom bolts. Now go to the top and drill two holes in the same pattern at 20 and 23 inches from the end.

Next, the floor joists should be installed. Install the second-floor joists first. Saw the 14-foot lumber to the correct length of 12 feet and 6 inches. Be sure to square the end before you start to measure. Once the correct size is obtained, measure back 3 inches on what will be the upper edges of the joist. Draw a line connecting this 3-inch mark with the lower end of the rafter. This will form the correct angle for the roof. Saw it off.

Now, with the upper floor joist correctly angled, you can fabricate the lower floor joist. Notice that the lower floor joist has a splice at the center (see Fig. 3-1). First cut two of the 12-foot 2 × 12's into 10-foot, 3-inch lengths. Keep the sections of the joist that were cut off since they will be used for the splice plates. Next, select one end of each to be used for the outer end and cut the ends to the correct roof angle. As in the upper-floor joist this first-floor joist will be cut back 3 inches at the upper edge.

When both planks have been cut to the proper size, lay them out on a level surface such as a floor or platform of lumber. Butt the ends together and stretch a chalk line from one end of the joist to the other to make sure both planks are lying straight. Then place the splice plates in position (see Fig. 3-1) and nail them with eight thirty-penny nails in each splice. Keep checking the joists as you nail them together to make sure the nails don't pull the boards out of the correct straight line. If this happens, the splice must be pulled apart and the joist renailed.

Once both the upper and lower joists are complete, the frame can be fabricated. First find a level place or prop the members up to make them level and layout the rafters so they are separated by 21 feet and 8 inches on the lower end and exactly 9 feet 6 inches on the upper end. This takes at least two workmen, three are better. Slide the 2 × 10 top floor joist into position between the rafters. The upper edge of the second floor joist should be 18 inches from the end of the rafter. Make sure everything is positioned correctly and then drill one of the bolt holes through the joist as indicated by the bolt hole in the rafter. Place a bolt in this hole and put the nut on the bolt but don't tighten it yet. Do this on each end. Next go to the first-floor joist and place it in position so it is exactly 8 feet 2 inches below the second floor joist, measuring vertically. Check the top and bottom of the rafters to make sure they are still positioned correctly and drill one hole in each end of the lower joist. If everything is right, proceed to drill the second hole in each top and bottom joist, install all the bolts, washers, and nuts, and tighten them up very well.

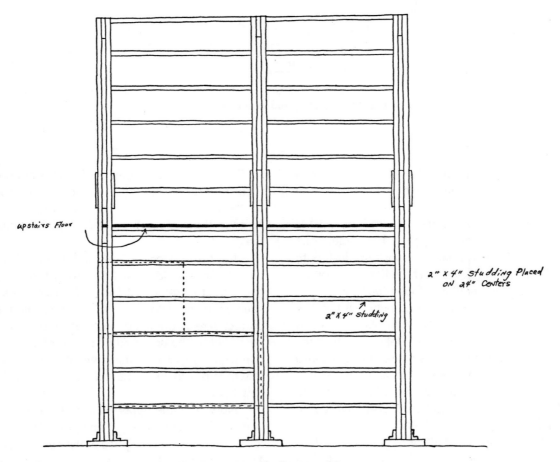

Fig. 3-3.   Partial side view of frame.

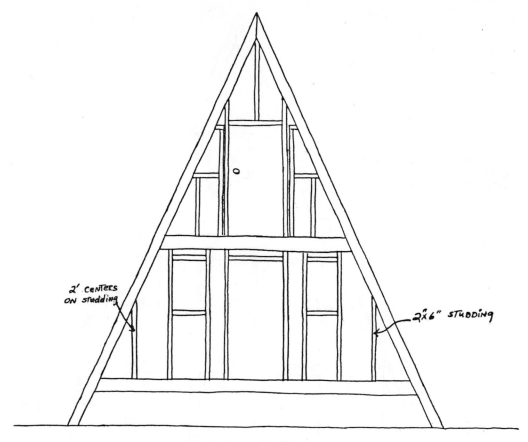

Fig. 3-4.   End wall frame.

With this done, drill the holes for the anchor plates. Then carry the frame to the proper position, prop it up in place, and install the anchor bolts. When this is done, one of the frames is complete and in position. Proceed to build the rest of the frames and put them up also. Join them together by tacking scrap board between the frames. When all twelve frames are fabricated, put in position, and plumbed up, the roof sheathing can be installed.

ROOF SHEATHING

If it is desirable to use planking for the roof, use tongue-and-groove 2 × 6's. A far less expensive way to cover the roof is with 5/8-inch exterior plywood. A problem develops here, though, because the rafters have 4 feet of space between them. This is too wide a space to bridge with 5/8-inch plywood; so additional framing members must be installed. The additional framing also helps when insulating the building. The framing members are 2 × 4's placed on 2-foot centers horizontally between the rafters. The 2 × 4's are nailed in position so they are flush with the upper surface of the rafters. Mark the locations for all of the cross framing by snapping a chalk line from one end of the building in the proper locations. As you will no doubt notice, this requires a lot of climbing. Temporarily nail some cross props between the rafters to make them sturdy and start nailing the cross braces in place. Start from the bottom and stand on the 2 × 4's to nail on the next one. (See Fig. 3-3.)

When all of the braces are in place, the first two rows of roof sheathing can be installed. The first row should be started 30 inches from the footings. Alternate the ends of the sheets. After the first two rows of sheathing are installed, it is practical to put the second floor in place. The second floor can then be

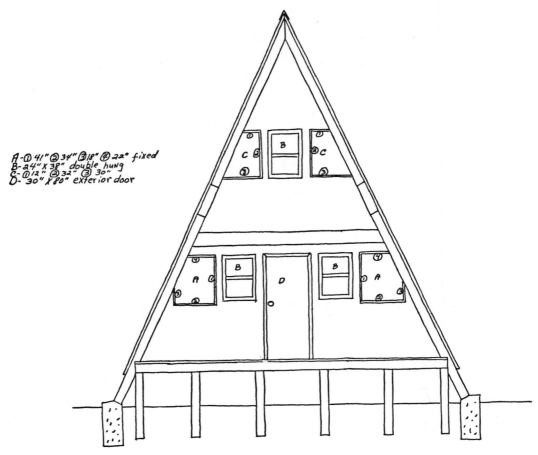

A-① 41" ② 34" ③ 18" ④ 22" fixed
B-24" x 38" double hung
C-① 12" ② 32" ③ 30"
D- 30" x 80" exterior door

Fig. 3-5. Front window and door sizes.

used as a platform to work from. We used conventional flooring of ⅝-inch plywood topped by ⅝-inch particle board. Again in this case the second floor could have been tongue-and-groove 2 × 6's or planking.

To make the floor more stable for conventional flooring, cross floor joists are installed on 16-inch centers. They are 2 × 8's nailed so they are flush with the top surface of the cross joist bolted to the frame. When two layers of flooring are installed, it is correct to lay one row in each direction to magnify the strength of each. Particle board, since it is very hard, is always installed uppermost on a floor. Actually, two layers of particle board can be used instead of plywood if desired. A 4-foot-square opening is left for the stairway.

Once the flooring is finished, the rest of the sheets can be installed on the roof. We found it advantageous to use a clamp and rope to pull the sheets up the side of the rafters. This takes two people. While one stands on the upstairs floor and pulls on the rope, the other works from a ladder and guides the sheets into the proper place to nail them. Be sure not to nail the plywood sheets on the roof so the edges all fall on one rafter. Alternate the lengths so each edge is 4 feet from the first. Nail them on with eight-penny coated nails. Be sure to drive each nail to the surface so the roofing will not be punctured later. Proceed to install all the sheeting on both sides.

MAKING AND INSTALLING THE RAFTERS

With the building erected to this point, it is advantageous to close it in as soon as possible. Do this by next fabricating the rafters for the upstairs roof. They are made as a unit and raised into place.

These frames must be made very carefully and

installed correctly or the peak of the roof will be out of alignment, which could cause all kinds of problems.

The following information covers building a typical frame. First select four straight 12-foot 2 × 6's from the lumber supply. Square the ends to make sure they are exactly 12 feet long. Then proceed to bevel the upper ends so they will fit together when assembled. The correct bevel is 10 inches. When all four rafters have been sawed correctly, cut out a 2 × 10 collar to fit in the peak between the four rafters (see Fig. 3-1). Further cut a 3-foot, 6-inch 2 × 4 tie piece to fit between the rafters. This will be located 5 feet below the extreme peak of the roof. When all of the pieces are sawed out correctly, lay out the butts of the rafters so they are separated by about 9 feet 6 inches. If any errors are present in your lower frame, the top frame will not fit; so it is well to "custom" make each upper rafter assembly to custom fit each lower rafter

assembly. Do this by measuring between the tops of the lower rafters at the inside edge. As mentioned previously, this dimension should be nearly 9 feet 6 inches. When the correct spacing is found, proceed to nail each rafter in its proper place to form the upstairs rafter assembly. Use the upstairs floor for a framing platform. (See Fig. 3-1.)

Installing the upper rafter assemblies so they are exactly aligned with the bottom rafter assemblies is somewhat worrisome. Care should be taken that the spread of the lower assembly fits the top of the lower rafter assembly when it is in place. Further, the upper assembly should be adjusted so that it is true with the building both crosswise and lengthwise. Fortunately both measurements can be determined by use of a plumb bob.

The first step is to find the exact center of the upstairs floor, measuring the full length of the building. When this is determined, snap a chalk line from one

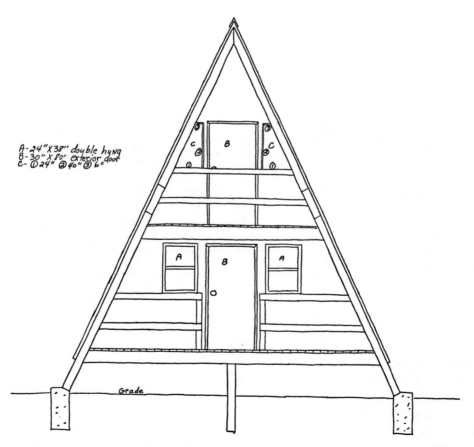

Fig. 3-6. Back window and door sizes.

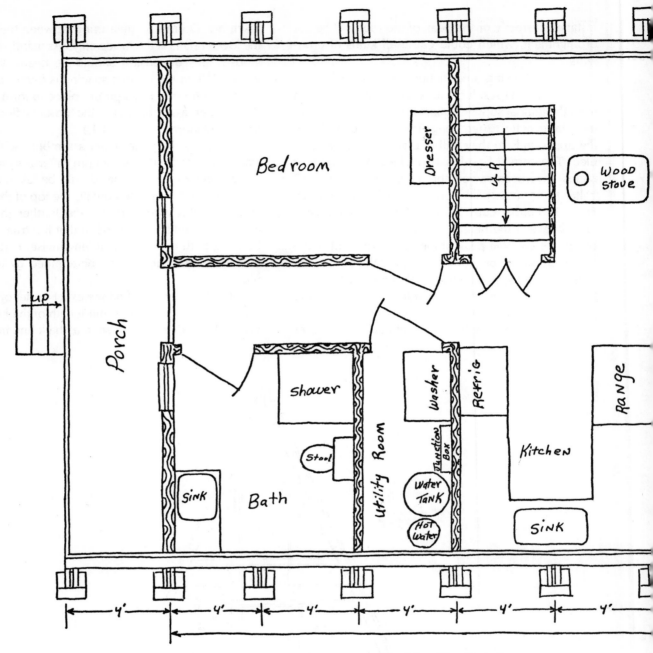

Fig. 3-7. First floor plan.

end to the other so a good clear line is visible. Next find the exact center of each lower rafter assembly and project this on the upstairs floor by snapping a chalk line across the floor between the centers of a pair of rafters. This will form a cross in the exact center of each rafter assembly. When this is done and

each upper rafter is completed and ready to be installed, set them up in the proper position. Affix a plumb bob to the exact center of the rafter assembly 2 × 4 tie piece and adjust the upper assembly so the plumb bob falls on the cross made by the chalk line. When the adjustment is correct, spike the upper

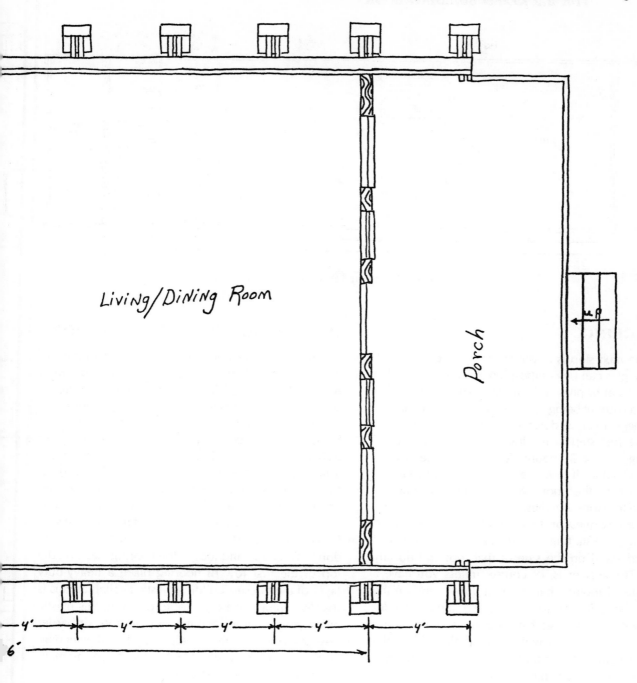

assembly to the lower assembly with eight twenty-penny spikes in each plate. Further, install a 1 inch by 6 inch by 2 foot splice plate on each side of the outside of the upper rafter assembly and nail it in place (see Fig. 3-1).

Repeat this same procedure for each of the assemblies. When the upper rafter assemblies are all complete, the roof sheathing can be resumed to close in the building so the interior will be protected from the rain. It is possible to work from the upper floor until the roof is almost closed in. The last panel has to be nailed on from the outside.

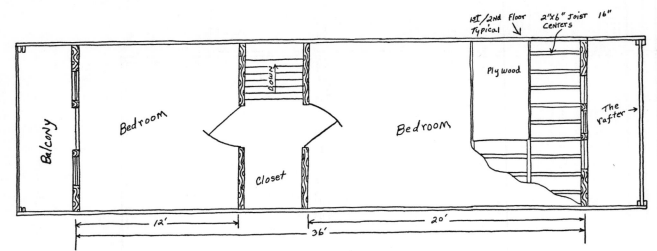

Fig. 3-8.   Second floor plan.

SHINGLING

The next step is to put the shingles on the roof. The shingles can be composition shingles or wooden shakes. Exterior plywood will start to deteriorate if left too long without being covered. Some contractors say it will begin to break down after three rainstorms.

The first step in roofing is to install a layer of building felt. If 30-pound felt is available, by all means use that. If not, use 15-pound felt. Install the strips across the roof, starting from the bottom. Overlap the strips 2 inches.

The first question that comes to anyone's mind when they are starting to roof an A-frame is how am I going to stand on the side of that roof and do any work? The answer is, of course, unless you have a professional roofer's harness you don't stand on the side of the roof. Instead you use two ladders, ladder jacks and a plank to go between the ladders. Ladder jacks are metal frames which can be placed over the rungs of a ladder to support a plank to stand on. They can be rented or purchased from commercial sources.

The roofer then works from the ladders. It is necessary to have someone below pass up the shingles, nails, etc. See chapter 1 for information on such things as how to install different kinds of roofing and take care of the flashings.

Once the roofing is complete, you can work inside snug and dry in almost any weather. Proceed to install the first floor next.

INSTALLING THE GROUND FLOOR

Install the first floor by nailing cross floor joists of 2 × 6 material between the 2 × 10 joists to form a bed for the layers of flooring. The 2 × 6 joists are installed on 16-inch centers as with the upper floor.

It is necessary to install a vapor barrier under the bottom floor since no basement is used with this building. After all the 2 × 6 joists are in place, staple a layer of builder's felt over the top of all the joists. Make sure the edges overlap and that no holes are poked in the felt as it is being installed. With this complete, put down the first layer of flooring. Be sure to use nails which lock themselves in place so they don't "walk up" and loosen the flooring. Now on the top of the first row of flooring place a heavy-duty layer of the thickest building plastic available. This is usually 6-mill. Take care not to puncture this plastic when you install it since any hole in a vapor barrier lessens its effectiveness. The plastic should extend up on the side of the walls at least a foot to permit making a good joint with the vapor barrier used on the walls. The last step is to install the second layer of flooring, which can be 5/8-inch wallboard. Nails should be placed in the flooring every 6 inches around the perimeter and on every cross joist.

END WALLS AND ROOM PARTITIONS

When the flooring is done, the end walls and room partitions can be installed. All partitions are made from 2 × 4 stock. The end wall studdings are located on 16-inch centerlines, the partitions studding on 24-inch centers.

MATERIALS LIST FOR INTERIOR OF A-FRAME HOUSE

1. 41 8-foot 2 × 4's for interior partitions
2. Approximately 50 4 × 8 sheets of paneling
3. 1728 square feet of insulation (6-inch fiberglass or 3-inch foam)

Doors and Windows
4. 3 exterior doors
5. Interior doors
6. 11 windows minimum

Miscellaneous
7. Wiring
8. Plumbing
9. Stove, furnace, or electric heaters
10. Trim for doors and windows, paint, stain, etc.
11. 504 square feet of ceiling tile or paneling
12. 1296 square feet of floor covering
13. Kitchen cupboards and sink
14. Bathroom sink, stool, shower, or tub
15. Bathroom medicine cabinet

The end wall studding is made from 2 × 6 material instead of the conventional 2 × 4 since it is important to use 6 inches of insulation. This is desirable in either hot or cold climates. A 30 × 80-inch door is used with a rough opening of 32 × 82 inches at each end of the building (see Figs. 3-5 and 3-6).

It is expected the utility room will be used for the cold and hot water storage tanks (see Fig. 3-7). It also is large enough to contain the washing machine and the electrical relay box along with several storage shelves for miscellaneous equipment.

The bathroom is large enough to include a tub if desired or a shower and tub both. The downstairs

Fig. 3-9.   Finished A-frame house.

bedroom contains 96 feet of floor space while the large bedroom upstairs covers 240 square feet and the small one 144 feet.

The living room is large enough for a family of four to six. The kitchen has plenty of space for full-size refrigerator and stove as well as most any built-in that you might desire.

PORCHES AND BALCONY

The interior should be insulated with 6 inches of insulation and covered with the paneling of your choice. First, however, a vapor barrier should be placed over the insulation inside the house. A porch is recommended at each end at the ground level while a balcony is built at the west end of the upstairs (see Fig. 3-8). Both porches and balcony can be built with 2 × 6 plank floors and railings. The foundation for the outer ends of the porches should be built with cedar posts, preferably treated.

The beams in the living room can be left exposed as the ceiling can be nailed to closures on the sides of the beams.

By all means bring the plumbing into the house in the closest place that is handy and run it along the edge of the building. The vent pipe for the septic system and the cold water line can be kept from freezing by placing an electric heating tape around it and wrapping the pipe and line with insulation. Be sure to provide some way to drain the pipes if the building will be left unheated in cold weather.

Even though this building is a very imposing structure when it is completed, there is nothing about building it that can't be done by anyone who can measure, saw, and nail. It will probably consume about six months of spare time work. The interior finishing may take even longer. We were able to build this home complete for about $6,700.00. A major insurance company appraised it at $27,000.00 when we were finished. The building costs are about a third higher now, but even so it is a very beneficial way to spend six months of your spare time. Also so far as I know the building inspector can find no way to criticize any part of this building.

Chapter 4

HARVEST KITCHEN AND ROOT CELLAR

Years ago it was a common practice to have a separate room or building for the housewife to use for canning the garden produce. This worked out very well since all the mess and confusion of canning was taken care of outside the living area of the house. This building was called a harvest kitchen, cool kitchen, or summer kitchen.

As time went by and housewives canned less and less and people became more migratory, the harvest kitchen gradually faded into obscurity. Now with home preserving of produce becoming tremendously popular again, it has once more taken its rightful place in rural and suburban building schemes.

Another pioneer or country innovation that has fallen into disuse is the root cellar. A root cellar is an excavation in the ground that is used to store crops in. It is almost ideal for storing root crops such as potatoes, cabbages, and rutabagas. It also can be used to store canned produce. Most root cellars are dug into hillsides where the drainage is good. Of course, if sandy soil can be found to dig in, this is almost ideal because loose soils such as sand promote drainage.

The root cellar in this project is a cellar underneath the harvest kitchen. This is not only convenient for storing crops directly from the garden but it is excellent for storing home canned vegetables and meat.

Selecting the site for erecting the harvest kitchen is most important. It should be in a well-drained place where the excavation for the root cellar will not create an unwanted swimming hole. Also, it should be convenient to the source of vegetables and as close as possible to the main house. Finally, roads both from the highway and from the fields should be convenient to the kitchen in order to minimize the carrying of heavy produce and canning supplies.

EXCAVATING FOR THE CELLAR

Start building the harvest kitchen by laying out and excavating for the cellar. If it is feasible to turn this part of the job over to a contractor, by all means do it. However, it is possible, if the soil is sandy to dig the excavation by hand. It should be 6 feet deep, 14 feet long, and 14 feet wide.

Since the wall of the cellar is 12 feet long and 12 feet wide, the extra width of the excavation provides room to install a drain tile if it is suspected that drainage will be a problem. It also provides room for working on the outside of the blocks so they can be sealed.

INSTALLING THE DRAIN TILE

The proper method of installing drain tile in an application of this sort is to build up the wall. Then lay about a foot of crushed gravel around the outside of the wall. Next place 6-inch drain tile completely around the wall about 6 inches away from its outside surface. The drain tile will have to terminate at a cistern or out on top of the ground at a level lower than the bottom of the wall. When the tile is placed, lay a covering of 15-pound roofing felt directly on the tile and then cover that with a foot or so of crushed gravel and continue backfilling as required. The purpose of the gravel drainage bed and the drain tile is to carry away the seepage before it penetrates the walls.

MATERIALS LIST FOR HARVEST KITCHEN

1. Approximately 400 8 × 8 × 16-inch concrete blocks
2. Approximately 3 yards coarse gravel
3. 2 8-foot and 2 12-foot 2 × 6's for sills
4. 12 8-foot 2 × 12's for floor joists
5. 6 8-foot 2 × 10's for bridging material
6. 6 4 × 8 sheets ⅝-inch particle board for flooring
7. 96 square feet plastic vapor barrier
8. 30 feet 2 × 12 boards for stair treads
9. 2 10-foot 2 × 12 boards for stairs strings
10. 40 8-foot 2 × 4's for wall studs
11. 6 12-foot 2 × 4's for extra wall bracing
12. 16 12-foot 2 × 4's for shoes and plates
13. 4 8-foot 2 × 4's for window and door rough openings
14. 6 8-foot 2 × 4's for gable studs
15. 1 14-foot 2 × 6 for ridgeboard
16. 18 8-foot 2 × 6's for rafters
17. 200 square feet of roof sheathing
18. 200 square feet of shingles or roll roofing. 5 pounds ½-inch roofing nails.
19. 350 square feet of siding
20. 6 windows, shutters, and 1 door
21. 20 pounds sixteen-penny nails
22. 10 pounds eight-penny common nails, miscellaneous nails as needed

The outside of the blocks should also be coated with a thick layer of tar, asphalt roofing cement, or a cement sealer. The inside can be sealed with concrete sealer paint.

POURING THE BASEMENT FLOOR AND FOOTINGS

However, before the block laying can begin, the floor and the footings of the basement must be poured. Lay out the footings by excavating a ditch 6 inches deep and 12 inches wide (see Fig. 4-1). The ditch will extend 4 inches outside the 12 × 12 perimeter of the wall. Next, level and tamp the floor of the basement. When this is done, pour a 4-inch layer of concrete over the floor. The footings are also filled level with the floor, which creates a 10-inch depth of concrete for the wall support.

When the wall and footings have cured for at least three days, the block laying can start. The method of block laying was covered in chapter 2.

If desired, an outside loading chute can be built into the cellar for unloading large amounts of potatoes or other produce. However, the cost is rather large and it does tend to be a source of trouble.

BUILDING THE BINS

It is suggested that all of the bins be built above the floor at least two inches to let air circulate under the potatoes and other produce in the cellar. This keeps them dry and tends to make them store better. Canned goods, of course, can be stored on shelves built against the walls. These can be fabricated in the basement or they can be built outside and brought in. Naturally, it will be more convenient to place them before the harvest kitchen is built overhead.

SILLS

Start building the kitchen by filling the cores of the top row of blocks with concrete. Set anchor bolts for the sills every 18 inches. Use 2 × 8 planks for the sills. They should be treated with a waterproof preservative.

WATER AND ELECTRICAL LINES

Also bring in a water line and electrical line through the basement. The water and electrical lines can come through the walls. Be sure they are sealed with tar around the entrance.

Sink Drain

Don't forget that a sink drain must be provided also. Except for a little detergent used to clean the jars and utensils no polluting effluent will come from the drain of this sink. Therefore a simple drainage system into the house septic or sewage system should take care of the water. It is also very possible just to run this drain from the sink into a cistern made by sinking a large barrel into the ground. A cistern also can be built with native stones or from cement blocks laid up without mortar. Of course, if a drainage tile is used to drain around the basement, the water from the sink can also be run into that tile. However, sanitary codes may prevent this in some places, and indeed it may be more convenient to simply run the drainage from the sink out on the top of the ground some place where it will not cause problems. This can be done by just attaching a garden hose to the sink. A 200-gallon holding tank would also handle the effluent.

WINDOWS

One characteristic of our harvest kitchen is plenty of windows. A bright, cheery interior is a definite plus when long hours must be spent canning produce. Moreover, windows cut down the need for artificial lighting. Window shutters are used with this building so they can be closed to keep out the sun. Shutters also make the building more secure from unwanted intruders when it goes for long periods of time without being used. Finally, shutters tend to hold heat in the building in winter, which helps the temperature in the root cellar below. Shutters should be painted to contrast, not blend with the exterior paint.

FLOORING

Once the sill is bolted in place, the floor joists can be installed. Since this is practically a 12-foot span, 2 × 12's should be used for floor joists, especially if it is desirable to keep the cellar uncluttered with support posts. The 2 × 12's additionally should be placed on 12-inch centers. Further bridging should be installed between the joists to stiffen them. Use only grade number one fir or pine joists.

In the front corner a 48 × 48-inch opening is left for the stairway. Two posts are installed at this point to brace the joists. Cross bridging or bridging should be used and installed every 4 feet. The flooring should consist of two layers of hardboard, 5/8 inch thick. Additionally, a vapor barrier layer of plastic should be placed between the layers of flooring. This keeps the warmth of the cellar down in the winter.

Nail the flooring on before the wall frames are put in position. If rains are a problem, cover the project with canvas or plastic to prevent the basement from filling up with water, which is difficult to remove. The floor should be one block above ground.

BUILDING THE WALLS

Once the flooring is complete, the wall frames can be made up and nailed in place. Be sure to have plenty of help when setting up the wall frames as they are heavy. Nail the shoes (bottom horizontal members) to the floor and into the floor joists with twenty-penny nails.

Start by building the front wall. It is important to check with your lumberyard to see what is available in window sizes and door sizes before you make the rough openings. Not all sizes are available in all places; however, sizes close to the dimensions given in Figure 4-2 should be used.

Both the shoe and the plate (top horizontal member) should be doubled. The double plate is shortened to make tying possible. The door frame rough opening is 30 × 81. Dimensions of the rough openings for the windows are shown in Figure 4-2. Finish the front wall frame and set it aside while the rest of the walls are being fabricated.

The side walls are symmetrical and have a large double-hung window (see Fig. 4-2). Notice also that the side walls are shorter than the end walls by the width of two 2 × 4's and that the top half of the plate and the bottom half of the shoe are the full 12-foot width so they reach over the bottom half of the plate and under the top half of the shoe on the end walls; this permits nailing (tying) together the plates on the end and side walls and the shoes on the end and side walls.

When the walls are all fabricated, set them up and nail them together. It will be necessary to set the front and rear frames up first and then raise the side walls one at a time into position and nail them to the applicable plate and shoe. The shoes also should be nailed to the joists.

CELLAR STAIRWAY

At about this time or even before, the stairway should be built for the cellar. Building stairways is not difficult even for the beginner if he has access to a framing square and level.

Start by making the "stringer," or the two pieces of lumber which form the sides of the stairs. This stairway is properly called a ship's ladder because of its construction and angle of departure from the floor above. The strings must be 10 feet long to give a proper "run" to the stairway so that it won't be too steep (see Fig. 4-3). They should be made from 2 ×

12 stock. Set them up temporarily in position and mark for the joist and floor angles. Do this by projecting the surfaces of the floor joist and the floor surfaces to the strings. This can be easily done with a level. However, mark both strings at the same time before you cut and make sure the level is deep enough into the stock to pick up a full cut on the board. Once the strings are cut to the proper length and angle, the treads or steps can be laid. They should also be made from 2 × 12 stock. The treads should be 40 inches wide to permit passing on the stairs or carrying wide loads up and down. An unorthodox, but very simple way to find the location for the treads on the string is to set the strings up in the proper position and temporarily tack the upper ends of the strings to the joist. Find the proper separation for the treads by laying a tread between the top and bottom of the strings. Then use the square to project the location for the top tread exactly perpendicular with the joist which it rests

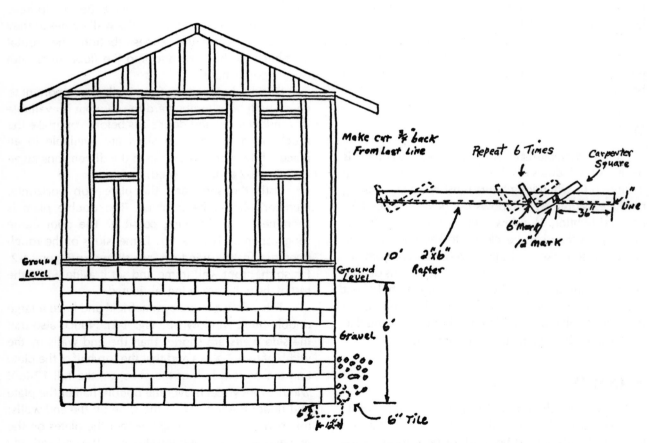

Fig. 4-1. *Left:* **Front frame and basement.** *Right:* **Making master rafter.**

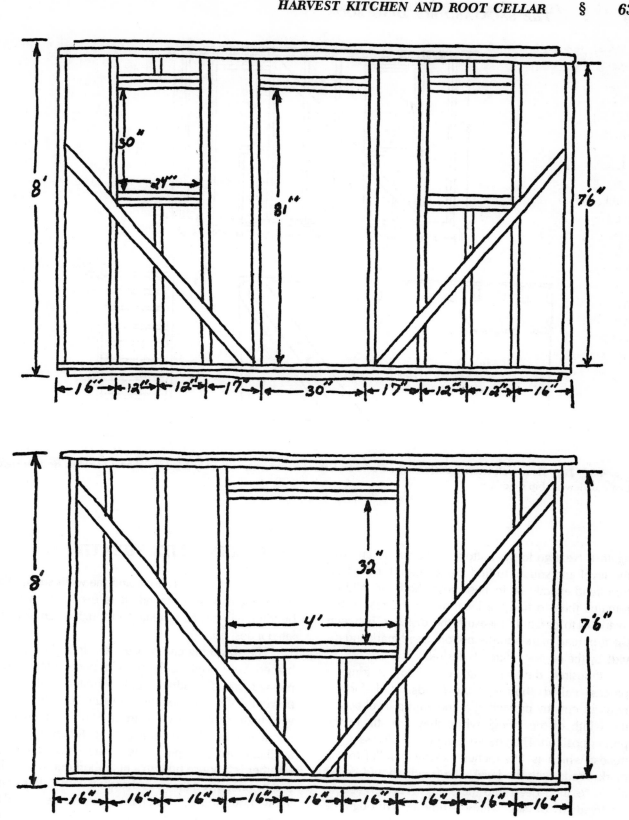

Fig. 4-2. *Top:* Front wall frame. *Bottom:* Side wall frame.

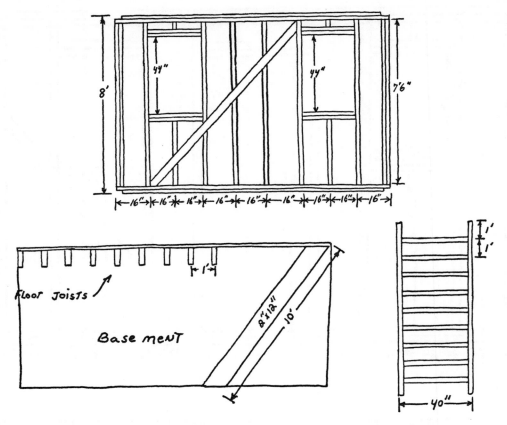

Fig. 4-3. *Top:* **Rear wall frame.** *Bottom left:* **Cross section of basement, with ship's ladder shown at right.** *Bottom right:* **Frontal view of ship's ladder.**

against. Next go to the bottom of the strings and use the level and square to find the location for the bottom tread exactly 10 inches above the floor. All that remains then to find the location for the rest of the treads is to carefully measure the distance between the top tread mark and the bottom tread mark and divide by the number of treads desired.

If desired, the treads can be set into the strings, a process called "housing" the treads. Mark for the housing cuts by projecting a line with a level across the width of the strings while they are still in the proper position. Then measure for the thickness of the treads, which is 1½ inches exactly parallel to the marks made with the level. The treads should be set about ½ inch deep into the strings. Further, secure the treads by driving twenty-penny nails through the sides of the string into the tread. If desired, cleats can be nailed under the treads to further brace them.

INSTALLING THE SHEATHING

After the ladder is done and the walls set up, the sheathing can be installed. It is expected that this building will not be insulated although it can be insulated if desired.

The sheathing can be Celotex since the frame is braced very well. Celotex is a panel made from pressed wood products and saturated with oil to prevent it from collecting moisture. Celotex has a very high insulating value in addition to being economical. Use ¾-inch-thick 4 × 8-foot panels if available. An example of the resistance of Celotex to weather is a house built in a neighborhood where the author once lived that was sheathed with Celotex and left without siding so that the Celotex was exposed to the air for about twelve years. No harm at all came to the Celotex.

The extra bracing which is necessary for using Celotex with a building can be installed anytime. It is merely an extra studding installed from the top to the bottom of the side walls in diagonal bias. One 14-foot 2 × 4 will do the job in each direction (see Fig. 4-2). This 2 × 4 is set into the existing studs or housed into the existing studs. Additionally, all door and window frames have an extra brace. Study Figure 4-2. If you do not desire to use Celotex sheathing, use ⅝-inch-thick exterior type plywood. Once the sheathing is finished, the rafter can be made up and installed.

LAYING OUT THE PATTERN RAFTER

Lay out the pattern rafter for the harvest kitchen roof by selecting a good straight 10-foot 2 × 6. Dress one edge to eliminate unevenness. Lay it on a pair of sawhorses with the dressed edge away from you. Measure and mark off a line the length of the 2 × 6, 1 inch from the dressed edge. Next mark off a point 36

inches from one end for the eave. Pick up the carpenter's square so the blade is in your right hand and the tongue in your left but with both sides of the square pointing away from your body. Lay the square on the measuring line so the 12-inch mark on the blade is exactly over the 36-inch mark for the eave (see Fig. 4-1). Further pivot the square so the 6-inch mark on the tongue is exactly over the measuring line. Take a scribe or a pencil and record this point. Also scribe a line where the blade of the square is located. This will be the plate mark. Next push the square to the left so the 12-inch mark on the blade falls exactly over the 6-inch mark on the measuring line that you just recorded. Notice that this moves the square to the left 13⁷⁄₁₆ inches. Since the span is 6 feet, we have to move the square six times to the left. After the sixth time mark the entire line where the tongue of the square crosses the 2 × 6. This line will designate the center of the ridgeboard. Next measure ¾ inch from this line to find the cutoff point for the rafter. Saw this off. Then go to the tail of the rafter and saw off the 1-

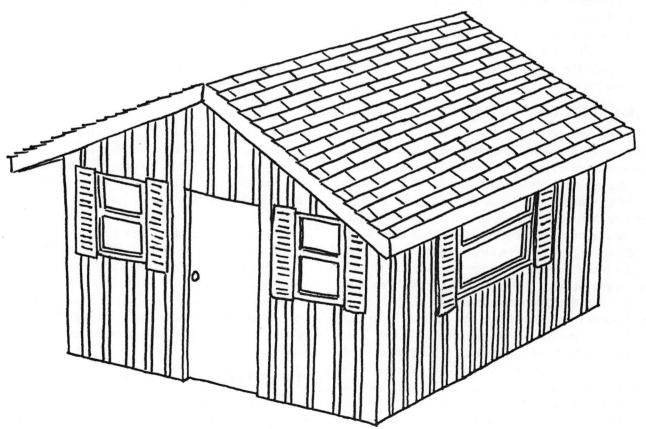

Fig. 4-4. Completed harvest kitchen.

inch mark to the point of the 12-inch mark for the eave. Then saw out for the top surface of the plate and the rafter is done. Make any slight adjustments needed and use this rafter for a pattern for the rest.

The rafters are installed on 16-inch centers. It will take nine pair of rafters. Toenail them to the ridgeboard with sixteen-penny nails, two on each side, and use rafter nails 6 inches long to nail the rafters to the wall plates. When they are all in place, the roofing can be placed.

The sheathing for the roof can be ⅝-inch plywood, 1-inch boards, or any one of the composition hardboards manufactured for sheathing. Cover the roof with asphalt shingles.

SIDING

Once the roofing is done, the siding can be placed. This building was planned with board-and-batten siding in mind. Eight-inch boards and 2-inch batten look good with this size structure. However, additional cross studs must be installed.

The harvest kitchen can be painted cabin brown and trimmed with apple green trim. Twelve-inch-wide fascia is utilized on the end of the eaves and the edges of the roof. The edges of the fascia can be scalloped. The top and bottom edges of the shutters can also be scalloped.

DOUBLE-COMPARTMENT SINK

This about completes the outside or shell of the kitchen. Inside the building we have to install a double-compartment sink. A reasonably priced, readily available unit of this type is the kind commonly called a laundry sink. Since these units are self-supporting, they will fit very well into the two-shelf cupboard which we must build in the kitchen anyway. In addition to the double-compartment sink a large draining area is utilized beside the sink. This can be self-made; it just consists of a three-sided box that drains into the compartment of the sink that finally drains out on the surface of the ground. Build this drainer by sawing out three pieces of 1 × 4. Two pieces are 3 feet long, one piece is 2 feet long. Nail the 2-foot piece across the ends of the other two boards and then saw out a panel of ⅜-inch particle board or similar water-resistant leakproof material. Incline this panel ¼ inch from the rear so the water will drain forward and out of the open end, and nail it

between the legs of the three-sided box just formed by nailing the 1 × 4's together. Further nail two 1 × 1 cleats along the inside of the box ½ inch above the bottom. The cleats will be level from front to back. Next, secure a 3 foot by 22½-inch mesh screen and use double-pointed tacks to secure the edges of the screen to the cleats. This forms a drainage screen for the washed vegetables. It is used by placing the washed vegetables from the sink on the mesh drying screen. The water which drains off the vegetables runs down the panel into the sink.

TABLE

Build your own table for the harvest kitchen also if you want. Although sometimes suitable tables can be found at flea markets and secondhand stores, usually obtaining one this large will be difficult. Start building by selecting a full ⅝-inch sheet of exterior plywood. Nail 46½-inch-long 2 × 4 cleats across the panel on edge. They are nailed one at each end and one in the middle. Use ten-penny finishing nails and space the nails 6 inches apart. Nail from the panel down into the cleats. Countersink the heads of the nails. Next nail a 1 × 8 across the ends of the 2 × 4 cleats at each side. Study Figure 4-5.

The next step is to install the legs and leg braces. The legs are also made from 2 × 4's. Cut six 2 × 4's 33½ inches long. Tip the table top upside down and nail the 2 × 4 legs so they are located in the corner formed by the legs and the 1 × 8 end pieces with the edge of the 2 × 4 against the crosspiece.

Use five sixteen-penny common nails at each joint. Further nail the 1 × 8 end pieces to the legs with three eight-penny nails at each leg. Now, the next step is to secure 1 × 8 boards and saw out braces for the legs. Each pair of legs has a double brace, one on each side of the legs. They are slanted from the top of the legs to 4 inches above the bottom of the legs. All projecting corners must be trimmed off.

The tabletop should be covered with formica or a tough smooth finished vinyl linoleum. The bare wood of the legs, etc., can be stained some nontoxic stain.

PLUMBING AND WIRING

This brings the harvest kitchen to the usable stage except that the plumbing and wiring must be brought in. If it is desirable to use an electric stove,

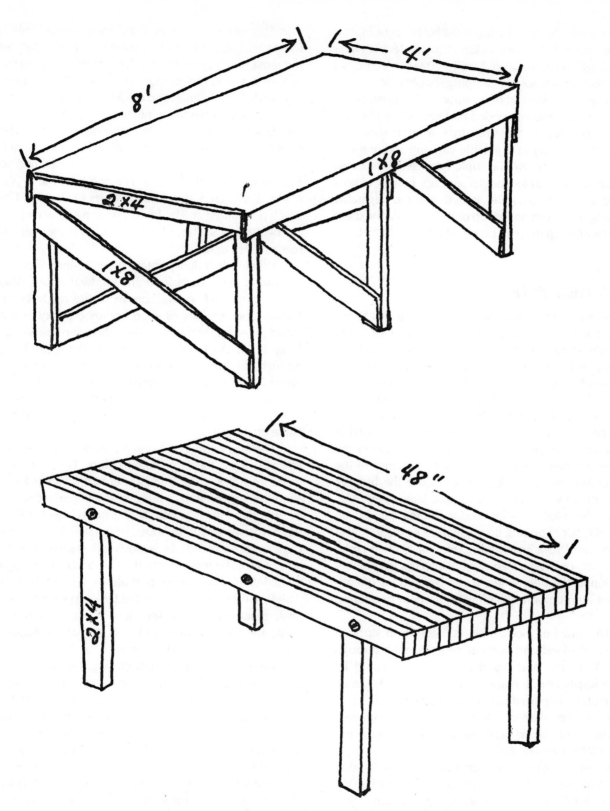

Fig. 4-5. *Top:* Harvest kitchen table. *Bottom:* Butcher block.

heavy-duty 220-amp wiring must be brought in; this is a job for a skilled electrician. A small electric water heater will be useful also. Small 110-volt electric hot water heaters are sold in camping supply places.

The plumbing can be as elaborate or as simple as you wish. It is recommended that you at least bring a cold water line to the sink since a large supply of water is necessary to vegetable canning and preparation. Drainage from the sink has been discussed earlier in this chapter, but to sum up, connect into a septic system if possible. If not, run the vegetable washing water out on the top of the ground. The drain water with soap in it can be piped into a holding tank.

BUTCHER BLOCK

Some of the equipment that is used inside the harvest kitchen can be homemade also. One good example is the butcher block. A butcher block, of course, was originally intended to be used by butchers for cutting up meat. It is very handy since it is sturdy enough to stand almost any type of hacking or sawing, and it provides a surface besides the table to work on. A butcher block has considerable aesthetic value also, and when it isn't in use in the harvest kitchen, it can be placed in a corner of the regular kitchen. Some people even use the butcher block for a TV stand or a stand for a large plant. Also the butcher block can be used in the regular kitchen every day for chopping up vegetables or meat. It also can be used as a snack table, regular kitchen table, card table, or even a step stool. It also is a very handy addition to the den, where it can be used for spreading out ammunition reloading equipment or a stamp collection.

This butcher block table is made from strips of hardwood glued and fastened together with through bolts. The very best wood for this table would be maple heartwood. If maple is not available, ash, oak, or one of the imported hardwoods can be used.

It requires sixteen 2 × 4's, 48 inches long, well seasoned and planed on three sides, to make the top for this table (see Fig. 4-5).

When all of the pieces are obtained, the first step is to drill them for the through bolts and to cut the notches for the legs. The holes are drilled correctly in all the pieces by first marking one of the 2 × 4's at 6 inches, 24 inches, and 42 inches. Carefully center the marks on the width of the 2 ×4 and then drill $7/16$-inch

diameter holes at each of these three points. This is the pattern and it should be used for drilling each of the remaining fifteen 2 × 4's. Do this by clamping the pattern 2 × 4 to the member yet to be drilled, centering the drill in the hole, and drilling through it.

The next step is to countersink the holes in two of the boards which will be on the outside. The countersunk hole is $3/4$ inch in diameter and $3/4$ inch deep. Further, two 2 × 4's have to be notched to receive the ends of the legs. The notch for the legs is made $2^1/2$ inches deep and $3^1/2$ inches wide. The notch is centered at 6 inches from each end of each of the two 2 × 4's. This, of course, cuts out the hole already drilled in it.

When this is done, obtain three 24-inch sections of $3/8$-inch ready bolts from any hardware store. Also get six nuts and six washers to fit the ready bolts. Further obtain a pint of good nontoxic epoxy or hot wood glue. Apply the glue liberally to the inside surfaces of all the layers of wood, slip the ready bolts through the holes, and tighten them to form the tabletop. Be sure the countersunk layer is on the outside and the layer that is notched for the legs is second from the outside. While the glue is hardening, make up the legs.

The legs for this table need not be hardwood unless you want them to match colors. They are also made from 2 × 4 stock and are cut 28.5 inches long. Sand the legs smooth to remove any imperfections in the wood. Fit the legs into the table only after the glue has hardened about 48 hours, or whatever the manufacturer recommends, by removing the two end ready bolts. Place the legs up into the slot, and drill through them. When this is done, coat all the sides of the legs which will be recessed into the tabletop with a good glue, place the legs in the slots, replace the ready bolts, and tighten again firmly. Let the table set again for about three days and it is ready to use.

When it is used for meat cutting, the top surface of the butcher block should be cleaned very well before it is put away. In fact, the top should be scraped so as to actually remove the top layer of wood so no meat juice or piece of meat adheres to the surface. Butcher blocks have reportedly been banned from butcher shops where they were used to cut meat for the retail trade because of the possibility of contamination of the surface by improper cleaning. This doesn't have to be a consideration if the surface is cleaned by scraping. The surface should be cleaned down periodically with soap and hot water also.

Chapter 5

GUEST HOUSE

My brother who lives in sunny California has an overflow of house guests from the Midwest during the frigid months. In fact, he said one morning he stepped on three abdomens and a larynx getting to the bathroom. He weighs 226 pounds, and the resulting shrieks of pain convinced him that he had to find some place for his surplus guests to sleep besides the hallway. They enthusiastically agreed.

At almost the same time a cousin who lives near a Michigan ski resort had a similar problem. He said people wanted to stop over and talk to him during the skiing season that he hadn't heard from since school days and he's forty-five. However, he loves skiing and was glad to hear from all these people. Naturally, during the peak of the season his visitors cannot find a place to stay since all the commercial accommodations are sold out far in advance. They all wind up staying at good ol' Cousin Albert's place and four to twelve extra people in a two-bedroom house soon leads to strained relationships, even among good skiing companions.

Both confided these problems to me during a get-together last summer. We talked it over and a possible solution developed. Build a small house or cabin just for the guests, a guest cabin as it were.

Some obvious problems were immediately apparent here, mostly centering around finances, since these good people were still paying the mortgages on their own houses. Total costs for the building could not exceed $1000, they mentioned through tight lips. Since low-cost buildings are a favorite subject of mine, I immediately began to search out structures that would answer their desires. They both live in the suburbs and the buildings would have to be adaptable to the neighborhood; yet they wanted something distinctive.

It didn't take me too long to settle on two designs that would have possibilities, and by the time the Yule season came around I decided I would have to visit them and explain how to build these. In fact, maybe even do the actual building. Building experimental buildings when someone else pays the cost is

Fig. 5-1. Geodesic dome guest house.

fun. My brother opted for a geodesic dome, style 2V ³/₈, no doubt being entranced by the beautiful sketch I made of it on a shopping bag with a red crayon. With him settled into gathering the materials, I tucked my sketch of the design I selected for Cousin Albert under my arm and immediately sped to Michigan. He and his wife both went into a trance when I showed them the Mongolian yurt I drew on real writing paper with a real pencil and explained to them just how it would be constructed. Grandmother O'Connor who was staying with them was interested also. She said it looked just like an uppity outhouse. Nevertheless we proceeded.

Now, I explained to Cousin Albert that a genuine Mongolian yurt is made from small flat boards similar to the boards that we call furring strips and they cover it with material similar to felt. In fact, we perused a book called *Build A Yurt* and carefully studied the baby gate design that author Len Chorney used for

building the walls where he held furring strips together with bent nails and tied them together with cable. "Too flimsy," said Cousin Albert. "Definitely too flimsy," said his building inspector. I didn't necessarily agree but I went home and decided to try to find something else to build it with that wouldn't cost any more but would be more acceptable. In the meantime my brother called; he said bring the family, come out for a short vacation, and let's start building the dome. We left immediately.

GEODESIC DOME

The first step in building the 16-foot geodesic dome was to select a location where the finished building couldn't block his view of the valley below and wouldn't interfer with tilling his garden or use of the lawn for gatherings or games.

MATERIALS LIST FOR GEODESIC DOME

1. 36 8 × 8 × 16-inch concrete blocks
2. 10 4 × 4's, 8 feet long
3. 120 feet 2 × 6 material for floor joists
4. 200 square feet 5/8-inch exterior plywood for flooring
5. 10 12-inch 2 × 4's for base hubs
6. 8 4 × 8 sheets 3/8-inch plywood for skin
7. 2 4 × 8 sheets 3/4-inch exterior plywood
8. 160 running feet of clear 2 × 4 lumber for struts
9. 40 square feet 1/4-inch-thick clear plastic for windows
10. Paint and roofing material as needed

Staking Out the Perimeter

When we had the site selected, we drove a stake in the approximate center, pounded a six-penny common nail into the top of the stake, tied a string to the nail, made a knot in it at 8 feet, and walked slowly around the center stake driving other stakes every foot to outline the perimeter. It took fifty stakes to outline the perimeter and even then we had one space between stakes which was longer then the others. No matter, we said to each other, since we only want ten equally spaced points around the perimeter anyway. To find the ten points we simply took the string and stake and used it like the legs of a giant compass to find ten equally spaced marks on the circumference of the circle. We drove red-tipped stakes at these points which outlined the foundation when they were all in place. These stakes marked the exact points where the foundation blocks would be set (see Fig. 5-2).

Setting the Foundation Blocks

Next we dug down at these points to an 18-inch depth and placed three 8 × 16-inch cement blocks one on top of the other in the holes. We used a string

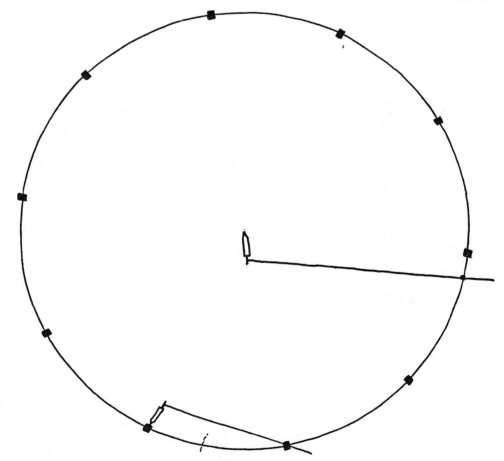

Fig. 5-2. Laying out the geodesic dome foundation.

and line level to make certain that the top surfaces of the top blocks were all even. Further, we went to the exact center of the foundation, which was the point where we had driven the original stake, and installed a double cement block foundation. Then we filled the cores of all blocks with cement. Finally, we set a 3/8 × 8-inch bolt in the wet concrete in the center of each perimeter piling. When this was done, we went to the mountains trout fishing for two days while they were curing. When we got back the concrete was hardened and we could continue.

Flooring

The next step was to purchase ten 8-foot 2 × 6's which had been pressure-treated with a creosote base, decay-preventive preservative. Then 1/2-inch diameter holes were drilled in the 2 × 6's, one hole in each piece so they could be placed over the bolts embedded in the concrete. These hole locations were found by laying the 2 × 6's alongside the bolts and projecting the bolt locations onto the planks with a pencil and square. With this done, the girders for the floor were in place. Next, 2 × 4's were nailed on 16-inch centers between the 2 × 6's (see Fig. 5-3). Then the flooring of 5/8-inch-thick exterior grade plywood was nailed on top of the girders and joists. Here we got into a slight argument. I wanted my brother to use tongue-and-groove 2 × 6's instead of the plywood, placed in a hex pattern which would have made a beautiful floor. I did finally get Cousin Albert to use it in his yurt, but that's getting ahead of my story.

Hubs and Struts

Now that the floor was done, we sat on it and traded tall stories for most of the next day while we were planning how to form the hubs for the dome

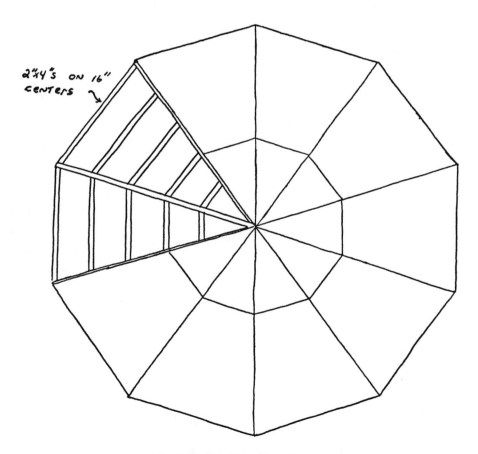

Fig. 5-3.   Geodesic dome floor.

Pent Hub

Hex Hub

Struts

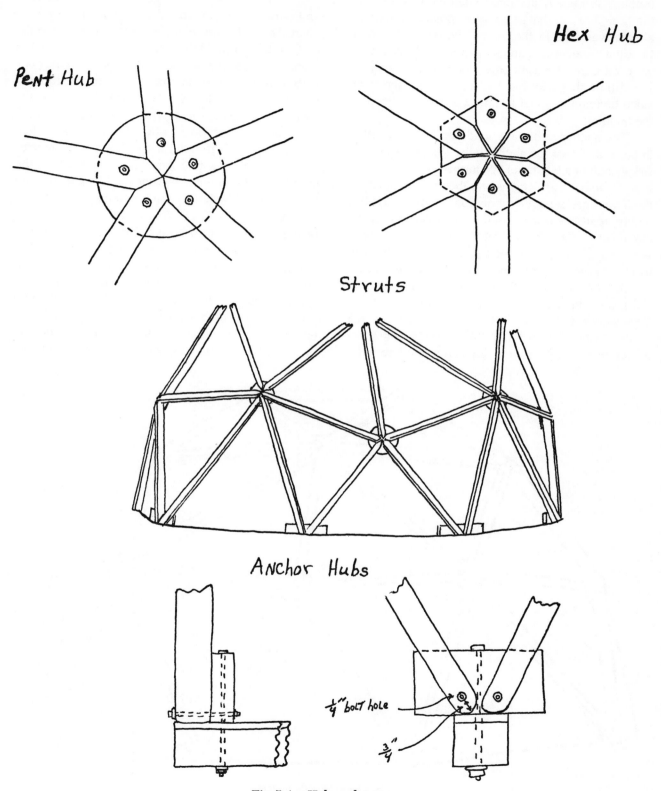

Anchor Hubs

¼" bolt hole

¾"

Fig. 5-4.   Hubs and struts.

framing. We finally decided to nail the base hubs to the floor and use that as a starting point. We ignored all the warnings in the various dome books about building a model first and started to work. When we were all done, I wasn't sure whether we gained or saved time that way but I suspect we gained time, since building a model is almost as hard as building the building.

The anchor hubs were made up first and nailed in position. It took ten anchor hubs. (An anchor hub is simply half of a hex hub.) We formed ours by sawing 2 × 6's into 1-foot lengths and then nailing them to the floor in ten separate evenly spaced locations 60½ inches apart, measuring on the circumference of the circle with a flexible steel tape. We didn't nail these very securely yet, since we decided they might have to be changed slightly after the struts were placed, which turned out to be the case. When all the anchor hubs were in position, we installed the first struts. They were made from 2 × 4's. It took ten 50½-inch 2 × 4's and ten 57½-inch 2 × 4's to complete the struts for the first row. They were fastened to the hubs with ¼-inch bolts, placed through the width of the strut and through the hub. This allowed them to be free to pivot. To hold them in position we just tightened the bolts so they gripped the struts. The correct angle was found by projecting half of a 4-inch circle on the hubs with a pattern. To make this pattern from cardboard draw a 4-inch circle on the cardboard. Then use the radius measurement to mark off six equally spaced circumference sections; lastly, cut the circle in half with a scissors. Naturally, a 4-inch plastic protractor would also fill the bill. Anyway, 4 inches from the center of the hub the centerlines of the struts should be separated 4 inches. At the center they should meet and rest on the floor. This requires that they be tapered at the end. Also note that one long and one short strut is fastened to each hub (see Fig. 5-4), but they alternate so that a short strut on one hub is adjacent to a short strut on the adjoining hub; likewise, the long struts are adjacent to each other on adjoining hubs. One way of thinking of this is to say each hub is either lefthand or righthand. In the correct positioning the lefthand hubs are all separated by righthand hubs.

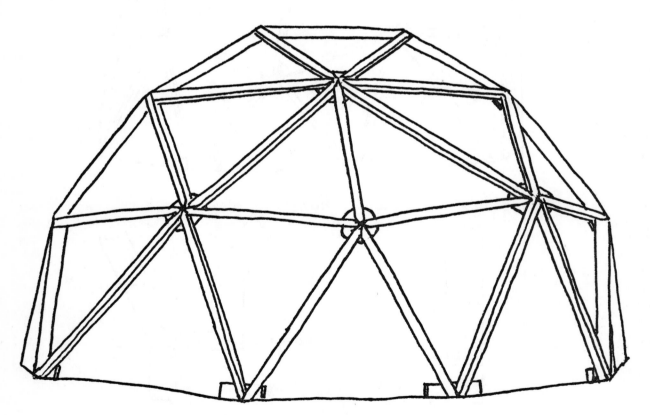

Fig. 5-5. Complete dome frame.

When we had more or less correctly positioned all of these hubs and had the struts in place, we noticed that the upper ends of the struts all nearly touched each other. This was indeed fortunate since the next step was to install the first row of full hubs (see Fig. 5-3).

All of the remaining hubs are either full hex or pent hubs. This structure, we found, required six pent hubs and ten hex hubs. We talked it over and decided to make them from ³/₄-inch plywood and bolt the struts to the hubs with ¹/₄-inch machine bolts. Accordingly, we purchased a sheet of ³/₄-inch exterior grade plywood, borrowed a compass, ruler, and pencil from one of my brother's daughters, and went to work. Since we needed ten pent hubs with 4-inch sides, we drew ten 4-inch diameter circles on the plywood and then, using the radius measurements of the 4-inch circle, we found six equally spaced points on the circumference of the circles and connected the lines with a straight edge. Presto. We had formed the outlines for ten hex hubs. We sawed them out with an electric saber saw and sanded the edges with a hand sander.

Next we drilled a ¹/₄-inch hole, recessed 1 inch from the points of the hexagon. When these were done, we set them aside and made up the six pentagon hubs. After tearing our hair, wringing our hands, and threatening numerous times to junk the whole project, we finally found that a 4-inch diameter circle can be changed into an almost perfect pentagon by making marks 4¹¹/₁₆ inches apart on its circumference and connecting the lines. Deciding that there might be a slight error here, we decided not to saw pentagon shapes out at all. Instead, we just cut out 4-inch circles and made pencil marks to indicate where the struts would radiate from the center of the circle. This worked so well that we wondered in retrospect why we had gone to the trouble of sawing out the hexagon shapes for the hexagon hubs. However, after the skin was put on, the combination of hexagons and circles had a pleasing effect. In fact, one of the first things we noticed was that the struts formed a hexagon where they joined at the hexagon hubs and a pentagon where they joined at the pentagon hubs. This too was aesthetically pleasing.

When it came time to put the hubs in position on the first row of struts, we had some trouble figuring exactly where they should go. Finally we developed a method which went like this. All round (pent) hubs had short struts joined to every point, and all hex hubs had long struts going to the six o'clock and twelve o'clock positions. After that it was easy. However, for this to work out, each long and short strut must be positioned correctly on the base hubs.

Once we had the whole framework joined together, we went over it again, adjusting the struts until all the triangles were symmetrical. When they were all in place, we drove eight-penny finishing nails through the hubs into the struts. After that was done, we found we could easily chin ourselves on the frame without affecting it a bit (see Fig. 5-5).

The complete frame was a source of wonder to us and we sat around admiring it and sampling good California wine. The neighborhood children were attracted to it immediately and began climbing all over it. I noticed the champion climber of them all was a little girl who hardly looked strong enough to climb on the seat of a bicycle. No doubt a dome frame would make a welcome addition to a playground.

Covering the Frame

After a while we couldn't think of any more reasons for not proceeding; so we began thinking about the covering for the frame. Since ³/₈-inch thickness exterior plywood was economical, easy to install, and readily available, we decided to use that. After studying the frame of our dome we decided we would have to cover each triangle separately since each triangle is at a slightly different attitude from every other. Simple enough, we agreed. We just measured each triangle, projected the dimensions on a sheet of plywood, and sawed them out. We also found each triangle was a slightly different size from every other; so each had to be carefully measured. Also, after the first one was made, we got the very bright idea of mortising the edges together, which eliminated the wide crack left at the top of right-angle edges. The angle we used for a mortise varied slightly, but was generally about 15 degrees. For windows we sawed out triangles of clear plastic and fitted them into the frame on opposite sides of the structure. My brother found out after he had used these windows for about a year that he couldn't keep the joints sealed; so he finally had a triangle window frame made up for it so that he could put a piece of glass in the frame.

When the frame was covered, we went over the entire skin, sealing all the seams between the joints with roofing tar. Then we were ready to start roofing.

Roofing

We had no illusions about this structure shedding rain as it was. Possibly it could have been fiberglassed to make a tight rainproof structure, but the price of this was rather forbidding. We also considered and discarded other sealing material such as asbestos roof covering, but in fact, couldn't come up with anything that was likely to beat commercial roofing.

Brother Gene finally found and used a type of asphalt shingle called Shangles. These shingles look like wood shakes but are easier to apply to the curvature of the dome than wooden shingles would be.

Installing the Door

Installing a door in a dome is simplicity in itself. You just take a piece of chalk and measure for the outline across the triangles. Make up a frame, brace the frame for the structure so it doesn't collapse when you cut out the struts, cut them out, put the frame in place, and nail them together. Before cutting, decide what sort of door you want. It, of course, has to be built out on the top so the frame will stand straight up.

When the frame was all put in place, we decided we should have just left one triangle for the door. This would have provided a unique door that retained the integrity of the frame.

When the dome was all done and we were standing back admiring it, the same little girl who was climbing on it before said, "How come you built an igloo, Mister? Huh, huh, what's the igloo for, Mister?"

A few days later the family and I left for Cousin Albert's to help him put up his yurt.

YURT

Cousin Albert, prudent soul that he is, had a model of a yurt from cardboard and was quite im-

pressed with it. He said that you should make a model of everything and he's probably right. Anyway, we laid out the foundation for his structure exactly the same way as we laid out the structure for the dome. That is, we selected the site and used a string and stake to draw around the perimeter of the circle. Then we staked it off. Things changed drastically then, however.

MATERIALS LIST FOR YURT

1. 12 10-foot poles, minimum 4-inch diameter, for foundation and side walls
2. 14 poles for floor joists, lengths to be measured in field
3. 140 square feet tongue-and-groove 2 × 6's for flooring
4. 140 square feet ⅝-inch Celotex for insulation under floor
5. 128 running feet of 1 × 4 furring strips to reinforce the Celotex
6. 200 running feet of 1 × 4 furring strips for sheathing base
7. 12 sheets of ⅜-inch plywood for sheathing
8. 96 running feet of 1 × 4 for batten strips
9. 2 windows and 1 door
10. 12 10-foot poles for rafters
11. 16 running feet of 2 × 6 for rafters reinforcement at the skylight
12. 200 square feet of roof sheathing, ⅝-inch plywood or equivalent
13. Shingles to cover the roof
14. 18-inch diameter ¼-inch acrylic plastic skylight
15. Lumber to partition the interior as desired
16. Interior furnishings, electric lines, plumbing, etc.

While we had been in California, Cousin Albert had been busy scheming. With his model as a guide he had developed a foundation frame combination based on wooden poles which are easily available in Michigan. After studying his sketch I wholeheartedly agreed with his ideas and we went to work.

Foundation and Side Wall Framing

The poles that we used for Cousin Albert's yurt had to be sunk in the ground 3 feet (see Fig. 5-6), extend 7 feet above the ground, and be no less than 4 inches in diameter on the small end. This, of course, took 10-foot poles that tapered very little. Fortunately, Cousin Albert had a country place with a stand of

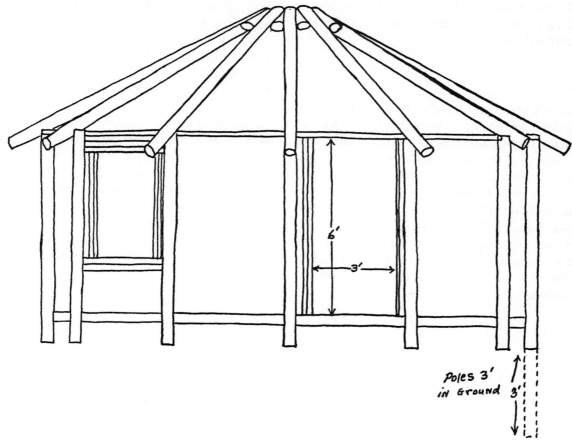

Poles 3'
in Ground 3'

Fig. 5-6. Yurt frame.

white cedar that could be utilized. We cut, peeled, let dry, and treated twelve of these poles before we started the actual building. The first 4 feet of the butt of the poles were treated with creosote; the remaining part with a commercial waterproof preservative. Probably if a person had to buy these poles, he would be better off to use something else belowground, such as concrete blocks. The postholes were back-filled with tamped concrete instead of dirt. We made sure each post was exactly straight up and down before the concrete hardened by laying a carpenter's level on two opposite sides of each pole. When the concrete did harden, we sawed all the posts off level at exactly 6 feet, 6 inches aboveground by using a chalk line and a line level to measure from one pattern post to all the rest. Once the correct level was found, we sawed each post off square by leveling and nailing a board on each side of the post and sawing directly above the boards.

Electricity and Plumbing

When this was all done we had the foundation and framing for the side walls complete. Next we had to bring the electricity and plumbing up to the floor. Working under the floor would be difficult since we left only a 6-inch crawl space.

We brought electricity to the yurt by tapping into the house entrance electrical box and bringing a new 30-amp, 110-volt circuit underground to our project. As an added precaution we installed a grounded fuse box in the yurt when we finished the wiring. The plumbing drains were connected to a 6-inch sewer pipe that was eventually connected to the house septic system above the septic tank. This, of course, posed no additional strain on the septic system since if the guests were in the main house they would be using practically the same amount of water which would end up in the same septic tank anyway. The water line was buried in the same trench with the electric line. When we had the work done which would have to be done under the floor, we proceeded to install the floor.

Floor

Since Albert had a woodlot that would yield an almost unlimited supply of free poles, we elected for economy's sake to use poles for the floor joists. The upper surfaces of the poles would be kept level by using a chain saw and chain saw guide to saw a flat surface on each pole. Another characteristic of our floor that made this possible was that we were going to use tongue-and-groove 2 × 6's for the finished flooring. Tongue-and-groove 2 × 6's can span 4 feet, so we would not have to place floor joists every 16 inches as is common. We used 4-inch minimum diameter peeled white cedar posts notched into the side wall

posts and nailed to them. We bridged from each side wall post to its opposite on the opposing wall, alternating the butts with the exception of the midcircumference posts, which were joined by poles placed at right angles to the rest (see Fig. 5-7). An additional row of short posts was placed at the centerline. As we installed each pair of poles we nailed 4 × 8 sheets of ⅝-inch-thick Celotex across the bottom surface of the poles. This had to be done as we went along since we couldn't crawl under the yurt after it was done. The Celotex was reinforced with 1 × 4 lumber furring strips placed across it at 4-foot intervals. We placed sheets of aluminum foil vapor barrier on top of the Celotex between the poles but we added no filler. Albert reports that the dead air space this produced was a very good insulator and the floor is never cold for his skiing guests.

Prospective builders who might not have access

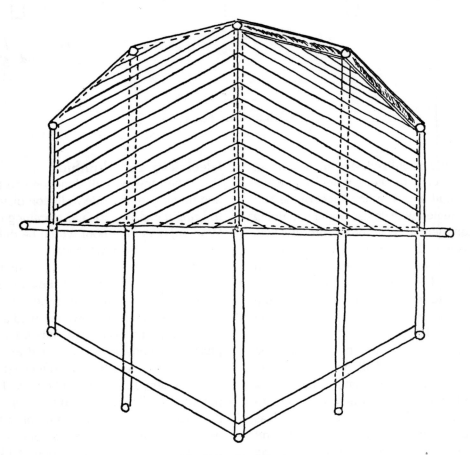

Fig. 5-7.   Yurt floor.

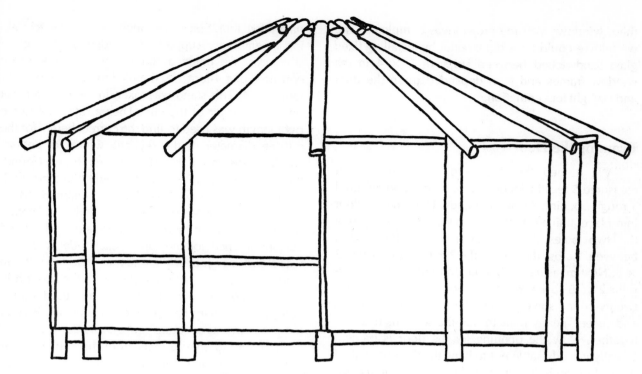

Fig. 5-8.   Yurt frame with sheathing.

to poles can make their floor joists from 2 × 8's or 4 × 4's and utilize the same spacing if they use tongue-and-groove 2-inch lumber for the flooring.

The 2 × 6's were put down in a hexagon pattern starting from the outside. When they were all in place, we nailed them down, sanded the upper surface, filled all the nicks, sanded the surface again, and put on a coat of wood filler. It would be varnished with three coats of a good deck varnish at a later date.

Sheathing

When the floor was in place, we started thinking about the side walls. Since the poles we used for the side walls were not quite evenly dimensioned, we had a choice of shaving or sawing the outer surfaces of the poles so they would be symmetrical or nailing milled lumber furring strips to them. This process commonly called "furring out" was what we eventually used. It consisted of nailing a furring strip to the center of the pole so that the sheeting could be nailed to a flat surface.

We considered several types of sheathing before

deciding on ³/₈-inch exterior grade plywood as the best for our purpose. Before we nailed the plywood to the poles, however, we lined the wall with 15-pound builder's felt, over-lapping the edges 3 inches, and stapling the felt to the poles after pulling it tight. See Figure 5-8.

Next we nailed on the sheathing, utilizing the horizontal method of nailing the sheets on with the full sheet on top. We used ³/₈-inch plywood since it is flexible enough to be bent around this radius. After the plywood was all nailed on, the joints were filled with caulking compound and 1 × 4 batten boards were nailed over the cracks. No additional siding was used; instead, the plywood and batten boards were painted with three coats of oil base cabin paint.

Windows

On the southwest and east walls windows were framed in with 2 × 4's. The windows we used had rough openings 2 feet 4 inches by 3 feet 6 inches. We used double-hung type windows with the sills, etc. already built for them. This saved considerable time but we were not fully satisfied with the appearance of these windows after the project was completed, and we felt we could have done better by installing fixed glass sandwiched between strips of 1 × 4 in the window frames and then depended upon the door and skylight for ventilation.

Framing the Door

We framed the door into the west wall. It measured 5 feet 11 inches by 36 inches and required a rough opening 72 by 38 inches. To frame the door into place, nail a 6-foot 2 × 6 to the inside surfaces of the two poles that the door would be located between. Across the tops of the 2 × 6's nail a 4-foot 2 × 6. Next measure the opening and place a third upright 2 × 6 to narrow the space to 37 inches. Use twenty-penny spikes to nail the 2 × 6's to the poles and sixteen-penny nails to nail the sections of 2 × 6 together. Now the opening for the door is framed. Usually it is advisable not to hang the door until all of the rest of the building, including bringing in the appliances, is complete. Knowing this, we left it out.

Installing the Roof Rafters

Once the above steps were completed, we had the building almost enclosed. The next step was to install the roof rafters. Again, since poles were easy to obtain, we decided to use poles for the rafters. Since we desired to have an opening in the center of the roof, which is conventional with the yurt, our poles did not need to reach clear to the center. In fact, after laying the angles out on paper we discovered that we wanted to have an 18-inch opening for the skylight and that we wanted the peak of the roof to be 10 feet from the floor. This would require the rafters to be 9 feet 7½ inches long. One foot of this length would be overhang in order to provide a wide eave, important to the overall appearance, and to prevent rain or snow from running down the side of the building from the roof.

Each rafter had to be custom-made, although we started with as near the same diameter poles as we could obtain. Using milled 2 × 4's, of course, would have been much simpler. The first step in making the roof was to find some way all the rafters could be held up on the unsupported ends until they would be nailed together. Since this roughly compared to making a temporary support for a ridgeboard in other structures, we made a stand from 2 × 4's which consisted of a 4-foot 2 × 4 nailed to one end of a 10-foot 2 × 4. This stand was taken in the door and set up in the center of the yurt. Then a rafter cut to the right length was used to find the correct height for the outside diameter of the skylight, which, of course, would be the end of the rafters. When this was found, the 10-foot upright 2 × 4 was sawed off to that length minus ⅜ inch. Next a section of ⅜-inch plywood measuring 20 inches square was nailed to the center of the 2 × 4. Then the rafters were laid in place and custom-notched for the top of each side pole. This was a two-man job; while one man held the rafter in place, the other marked the profile of the post on it. Then we took the rafter down again and used a small bow saw and sharp hatchet to cut out the notch. After the notch was as accurate as we could get it, we used a chalk line to snap a mark from the center of this notch to the center opposite end of the pole. This mark was then used as a guideline for beveling the end of the rafter. This bevel was set at a 30-degree angle since each rafter makes up 1/12 of the 18-inch diameter skylight. As the rafters were fabricated, they were laid in place and evenly spaced. Wedges were then custom-fitted and nailed into place between the rafters. The last rafter had to be custom-fitted since all the errors were apparent here. It was toenailed into each of its mating rafters and wedges. See Figure 5-9.

When all the rafters were securely in place, we removed the stand. Then in sheer exuberance I chinned myself in the skylight opening and, even though the ends were totally unsupported, they didn't even creak.

Roof Sheathing

We spent a few days thinking about what would make a good material for roof sheathing. We discarded the notion of using ⅜-inch exterior grade plywood since forming the sheets to the conical roof would be difficult and wasteful unless we first ripped the sheets into narrow strips. We had about decided to use 1-inch boards in whatever widths we could get them in, and, in fact, were talking to the owner of a small sawmill about buying some unplaned boards

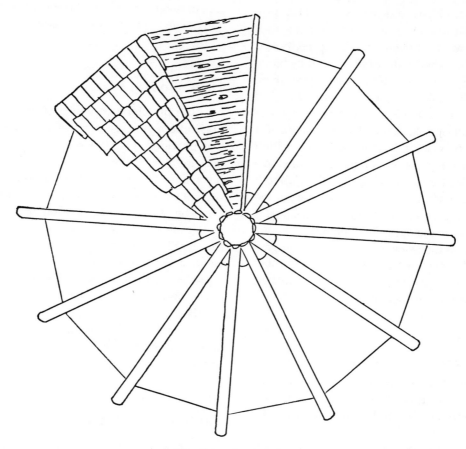

Fig. 5-9. Yurt roof.

from him when I happened to notice a huge pile of slabs he had lying there. Twenty minutes later we drove out with a pickup load of slabs, costing the magnificent sum of $1.00 and more than enough to cover the roof. When we got home we ripped the slabs into even widths with Cousin Albert's table saw, peeled the bark from them, and treated them on all sides with Penta. This would kill all the worms and fungus which were eating up the wood of the slabs. Then we laid them together as tight as we could get them and nailed them to the rafters. When it was done, we agreed that this was as good a roof sheeting as we could obtain even if we had paid out a lot of money.

Shingling

However, we still had to cover the roof to make it rainproof. Since it would seem a sacrilege to cover

this slab roof with anything artificial, we decided to use wooden shake shingles. Further, Albert decided we could split them out ourselves from two huge old cedar trees that he had in his woodlot. I was slightly taken aback by this suggestion but decided to go along with him. After all, I could always leave if the going got too rough.

Splitting shingles turned out to be more fun than work even though it took us three days of more or less steady labor. First we cut down two of the biggest cedar trees he had and sawed the trunks into 2-foot lengths, being very careful to saw them straight. Then we shaved the bark from the outer surfaces of the trunks and, using our wedges, split them exactly in half. Next, we chopped out the heartwood by splitting across the grain. All that remained then was to split each one of the sections into sections that were 1 inch thick at the circumference and tapered more or less to a point at the center. We started out using a chisel to split the pieces and ended up with a froe made from a drawknife, which worked much easier. Our shingles weren't all even; in fact, some looked like boards but they almost all could be used by fitting one into the other. We also noticed one good thing about homemade shake shingles; no one can tell if and where you made a mistake roofing. We overlapped the shingles at least ½ of their length. If a shingle was too thick at the lead end and we couldn't fit it anywhere, we tapered it with the table saw. We nailed the shingles to each other and to the slab roof with zinc-coated eight-penny nails. Fortunately, we had nailed the slabs on across the roof so the shingles could be nailed on in the opposite direction.

Since cedar shingles age naturally to a soft gray color, we decided not to use any finish. Any other wood could be used for shingles also if it were treated with a good waterproof preservative.

How to Waterproof the Skylight

Making the skylight properly waterproof required a special procedure. Directly around the rim we built up a section with 2 × 2's. On top of this section we installed a 20-inch diameter, ¼-inch-thick acrylic plastic disc and held it in place with screen bead. All exposed wood was painted with a good waterproof preservative. Further, the area around the ¼-inch-thick acrylic plastic was sealed with caulking compound. Albert reports this skylight "has not shed a tear since we put her up."

Wiring

With our outside problems pretty well solved, we decided to finish the inside, concentrating on the electricity and plumbing first. Wiring the inside was very simple. We divided the entire house into two circuits, one for the outlets and one for the lamps. Both circuits were protected by a 15-amp circuit breaker back in the house. The overhead light in the living room was controlled by a wall switch near the door. The bathroom lamp was turned on and off with a pull chain. No lights were installed in the bedroom since it was anticipated that dresser lamps would be plugged into the wall receptacles (outlets). A receptacle

was provided in each bedroom, one in the bathroom, and one in the living room to power a TV or radio.

Heat and Hot Water

The plumbing consisted of running a cold water line in from the master house plumbing. No hot water was provided since it was expected the guests would heat the water for shaving, washing, coffee, etc. on a hot plate. In winter when the wood stove was being used, a teakettle of water could be heated on that.

The stove was a logwood heater set up in the center near the bedroom walls (see Fig. 5-10). A freestanding fireplace also could be used with this setup if desired. At night a 110-volt electric heater equipped with a fan is used in very severe weather since this type of wood stove will not generally hold a good fire overnight.

The interior, including the ceiling, was insulated with 4 inches of foam insulation placed between the poles.

Custom-made Interior Siding

Since Albert wanted to maintain a rustic interior, we decided to nail furring strips vertically to the sides of the poles so that about ⅓ of the diameter of the poles would still be visible inside the rooms. Then we installed half-log siding that we made ourselves from aspen slabs.

Our custom-made aspen siding required considerable work. First we found a sawmill operator who was sawing aspen logs and visited him every day, carefully selecting slabs that were about 1½ inches thick and from 4 to 6 inches wide. We trucked these home and peeled the bark from them. Then we used Albert's table saw to saw out alternate 4- and 6-inch-wide siding boards. As fast as we had them sawed, we cut them into the proper lengths and nailed them in place between the interior logs. Then we immediately applied a thick coat of varnish so they would not darken, since untreated aspenwood tends to look dingy after aging. Making the siding turned out to be the most tedious job of all and I doubt that we were well repaid for our time, although it is certainly unique. Paneling or commercial exterior

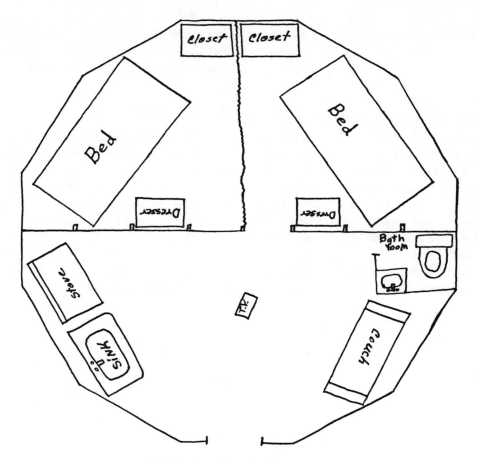

Fig. 5-10. Floor plan for yurt.

Fig. 5-11. Completed yurt.

siding would have been much faster and easier to use.

Inside Walls

We made the interior partitions from a single thickness of 3/8-inch-thick plywood nailed to 2 × 4 studding. The studding was placed with its width parallel to the plywood and was exposed inside the bedroom. No swinging doors were installed in the partitions; instead, privacy was maintained by a burlap curtain which slid on an overhead wire. The panel behind the stove was made fireproof by building a 33 × 72-inch section of brick wall with the width of the bricks placed against the wall. The brick was fastened to the partition wall for stability by stapl-

ing loops of wire to the wall so that the loops fell between the brick joints, where they would be incorporated into the mortar of the joints.

Outside Entry Door

Cousin Albert purchased an exterior type of door called a garage entry door for the main entry door to his yurt. This pleased me no end since I thought he might want to saw down a tree and hew that out too. We hung the door in the 2 × 6 frame, installed the hardware, and the yurt was done.

We left the next day, slightly under the weather from too much of Albert's homemade wine at the christening but pleased nevertheless at the guest house I had helped build.

Chapter 6

A WINDMILL, A WIND GENERATOR, AND A WIND MOTOR

WINDMILL

A smoothly rotating replica of a Dutch windmill on the lawn adds grace and charm to the most decorative residence. Moreover, it will transform a drab corner of the yard or garden into an eye pleaser that will delight the homeowner. Almost anyone who has a few tools can build one.

MATERIALS LIST FOR WINDMILL

1. 4 6 × 21-inch pieces of 1/8-inch masonite for propeller blades
2. 4 15-inch lengths of 1-inch diameter dowel
3. 6-inch diameter, 3-inch-thick wooden hub
4. 2 running feet 1/2-inch diameter plastic water pipe
5. 4-inch length of 4 × 4-inch lumber
6. 28-inch 2 × 4-inch board
7. 9 running feet of 2 × 2-inch boards
8. 2 × 2-foot section of sheet metal for canopy
9. sheet 3/8-inch exterior plywood
10. 1/2 × 10-inch machine bolt / 3 nuts / 4 washers

Start by building the four-blade propeller. The first step is to make the vanes. Procure a sheet of 1/8-inch masonite. Saw off four pieces measuring 6 × 21 inches. Further saw out four pieces of 1 × 2 lumber 15 inches long. Use a drawknife, spoke shave, plane or lathe to create a 1-inch round dowel 3 inches long on one end of each piece of lumber.

Center the masonite blades on the 1 × 2 pieces and fasten them together with wood screws so the inboard ends are flush with the dowels. See Figure 6-2.

Some people find that they like latticework propellers better. To make them requires four 2-foot 1 × 2's, nine 6-inch lattice strips, and four 24-inch lattice strips. As with the masonite propeller, dowel 3 inches of one end and then nail the lattice strips to the 1 × 2 strips as shown in Figure 6-2.

Next saw off a 3-inch hub from the end of a 6-inch diameter cedar or hardwood post. A section of 2 × 8 will work if no round stock is available. Carefully find the center of this hub and drill a 3/4-inch hole

Fig. 6-1. Windmill.

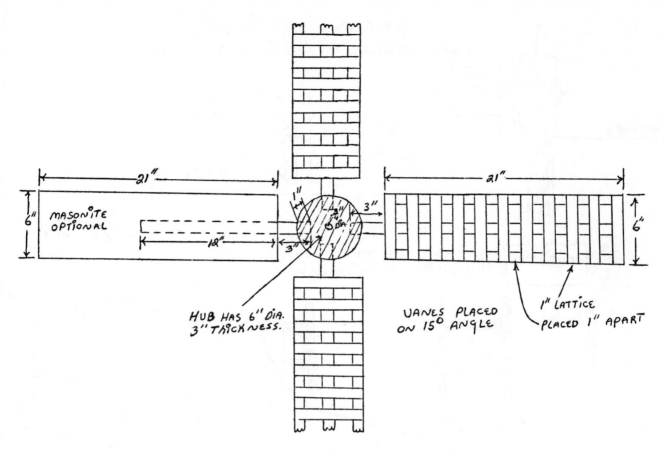

Fig. 6-2.   Four-blade propeller.

completely through the piece. This is the pivot hole and will receive the axle. Next drill four 1-inch holes evenly spaced and centered at the circumference of the hub. These holes should be 1 inch deep. Now fit the dowels of the propeller blades into the hub and rotate them so they are 15 degrees from the perpendicular and parallel with each other. Study Figure 6-2. When you are certain they are positioned correctly, drive a #5 finishing nail through the hub to hold them in place. If any holes are oversized so the dowel is loose, it should be glued as well as nailed. With that done, the propeller is ready to fasten to the axle.

Obtain a 2-inch-long section of ½-inch diameter plastic water pipe. Press one end into the hole in the center of the propeller. If it doesn't fit securely, coat it with a good wood glue before it is pushed into position. Let it dry at least eight hours before mounting it. Next, procure a ½-inch machine bolt 8 inches long. Also get three nuts and three flat washers for this bolt. If the bolt has more than 2 inches of unthreaded

shank, have the threads cut to within 2 inches of the head. Refer to Figure 6-3.

Try the bolt inside the plastic pipe. If it fits very tightly, ream the pipe out with a ½-inch drill so the bolt will rotate smoothly. When it does turn easily, place a washer on the bolt and slide the threaded end through the propeller hub so the head of the bolt is on the outboard side of the propeller. At the inboard surface of the propeller place another flat washer. Then thread a nut on the bolt up to the end of the threads and back it up with another nut. Turn the nuts against each other to jam them. The bolt should turn easily in the hub of the propeller.

Next, cut a piece of 4 × 4, 4 inches long. Drill a half-inch hole at its center. Nail it on the width of one end of a section of 2 × 4, 28 inches long. Twelve inches from the opposite end of the 2 × 4 drill a ¾-inch hole through the center of the width of the 2 × 4. Then bolt the propeller through the ½-inch hole in the section of 4 × 4. Next, take a 6-inch-long section

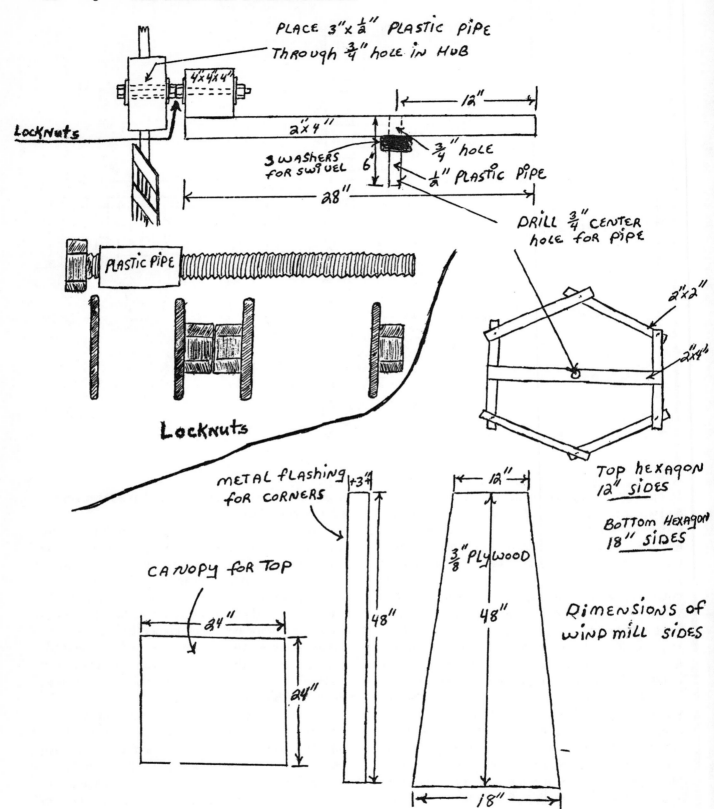

PLACE 3" x ½" PLASTIC PIPE
THROUGH ¾" HOLE IN HUB

4" x 4" x 4"

2" x 4"

12"

Locknuts

3 WASHERS FOR SWIVEL

¾" HOLE

½" PLASTIC PIPE

6"

28"

DRILL ¾" CENTER HOLE FOR PIPE

PLASTIC PIPE

Locknuts

2" x 2"

2" x 4"

TOP HEXAGON 12" SIDES

BOTTOM HEXAGON 18" SIDES

METAL FLASHING FOR CORNERS

+3"

48"

12"

3/8" PLYWOOD

48"

DIMENSIONS OF WINDMILL SIDES

CANOPY FOR TOP

24"

24"

18"

Fig. 6-3.

of ½-inch plastic pipe and glue it into the ¾-inch hole in the 2 × 4 so the end is flush with the top surface of the 2 × 4.

The base is framed from two hexagon figures made from 2 × 2 stock. It is covered with ¼-inch plywood. Cut six 18-inch 2 × 2's and six 12-inch 2 × 2's. Lay out the 18-inch pieces to form a hexagon approximately 3 feet in width. The ends can be lapped and one nail placed at each joint. Move the pieces until an even figure is produced; then place another nail at each joint to hold them in place. Further, nail a piece of 1 × 4 stock across the figure to brace it. Do the same for the 2-foot hexagon and brace it with a 2 × 4 nailed across it. At the center of this 2 × 4 drill a ¾-inch hole and ream it out so the ½-inch water pipe will rotate easily in its bore. Saw off any projecting corners.

Next, saw six pieces of ⅜-inch exterior plywood into sections measuring 48 × 6 × 16 inches. They are the sides of the windmill. Use them to nail the hexagon figures together forming a stand 48 inches high.

Then take the previously fabricated propeller and propeller mount, slide 3 ¾-inch washers over the plastic pipe, and push the pipe down into the ¾-inch hole in the 2 × 4 that is the brace for the 2-foot hexagon. This should create a smoothly operating pivot joint for the propeller. Its purpose is to allow the propeller to orient itself to the wind. Study Figure 6-3.

Next, find a piece of metal roof flashing and cut out one piece measuring 24 × 24 inches and six pieces measuring 3 × 48 inches. Nail the 24-inch piece across the top to form a canopy for the propeller mount, and use the 3-inch pieces for covering the corners of the base. It will be necessary to toenail a 1 × 4, 24 inches long, on edge to the center of the propeller mount to fasten the canopy to.

For realism cut windows in the base and construct a platform around the bottom (see Fig. 6-1).

Now, no doubt, after you have had your Dutch windmill in operation for some time you will begin to wonder if you can't use some of the wind power to actually do some kind of work. Perhaps like us you would like to light a light from your very own home-built wind generator. Now that we can no longer take our energy sources for granted, government and well-financed institutions of all kinds are frantically developing alternate energy sources and wind power

is one of the most worked with sources. However, if history repeats itself, it will still remain for lonely inventors working with hardly any resources but their own imagination and initiative who will develop the mechanisms and ideas that will carry these projects to the point where they are actually feasible. Perhaps you or one of your children will be the genius who does. Also perhaps this next project, though simple, will stimulate your mind along these lines. At the very least it will deliver much personal satisfaction when it is finished.

WIND GENERATOR

The next project is a wind generator that will light 12-volt lights or trickle-charge small batteries. It also has the potential to provide enough current to take care of minimal lighting at remote cabins or workshops.

MATERIALS LIST FOR WIND GENERATOR

1. 6-foot 2 × 6, white spruce or the equivalent
2. 4 × 8 sheet of ¼-inch tempered masonite
3. 2 8-foot 2 × 4's
4. 6-foot 4 × 4
5. 3 × 3-inch pipe nipple. ¾-inch pipe nipple. ¾ × ⅛-inch pipe nipple. ⅛-inch pipe nipple. 3 ¾-inch pipe tee's.
6. Bicycle tire, bicycle generator, bolts to hold tire to propeller

The first step in building this model is to obtain a 2 × 6-inch board, 6 feet long, that can be used for a propeller. Use white spruce if it can be found, but redwood, white pine, or similar wood that is straight-grained and knotfree can be used. The 2 × 6 should be well seasoned so it will not warp after it is put up. See Figure 6-4.

Place the 2 × 6 on the work bench and find the exact center of the piece. This is done by drawing a centerline both lengthwise and crosswise. Where the lines intersect, drill a ¼-inch hole. The next step is to mark off and cut out the material that must be removed to taper the width of the blade. Measure out 4 inches on each side of the center hole. Draw lines across the 2 × 6. Then go to the ends. At each end mark off 4 inches of width, measuring from opposite edges. Connect the end marks to the center marks at

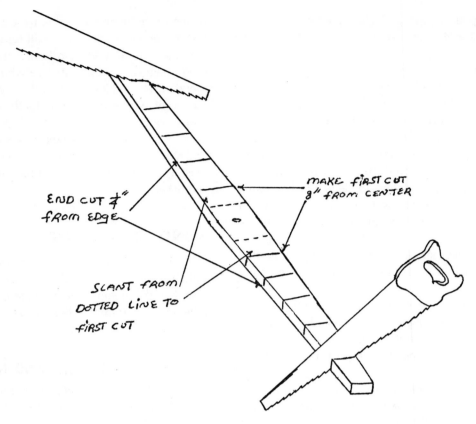

Fig. 6-4.

the junction of the center marks and the outside edge of the piece. Notice that this sections off a triangular piece 4 inches wide at the base with a height of 32 inches. Also check to be sure the triangles oppose each other. When you are sure you are removing material from opposite sides of the piece, saw out the marked triangles. With the width of the piece facing you, the edge of the propeller that hasn't been sawed off become the leading edge (see Fig. 6-5). Mark this edge for future reference.

With the width of the blade tapered, the next step is to form the thickness of the blade into an airfoil design tapering from ³/₄-inch thickness at the ends to the full 1¹/₂-inch thickness of the original board at the hub. Lay this out by drawing lines across the width 8 inches on either side of the center hole. Now take a hand saw and make diagonal cuts approximately ³/₄ inch apart from the 8-inch marks to each end of the piece. Caution: before making any saw cuts study

Figure 6-4. These saw cuts are to make it easier to remove the material which must be removed to form the flat side of the propeller; be sure to stop them about ¹/₄ inch from the edges to prevent too much material from being removed. After the saw cuts are made, remove the material between the cuts with a wood chisel or drawknife. Use a drawknife to smooth the flat surface. Clamp the propeller in a vise or to a bench with **C** clamps to make it easier to work on. Also use the drawknife to taper the hub into the flat side of the blades. Now the blade is well on the way.

To finish the blade in the airfoil design, turn it over so the flat side is down and shave the corners with a drawknife (see Fig. 6-6). Some people make patterns to keep track of their progress as they shave the back of the blade, but it isn't really necessary. Just work slowly and carefully and remember that the propeller will turn clockwise and the thickest part of the airfoil design will be the leading edge. The trailing

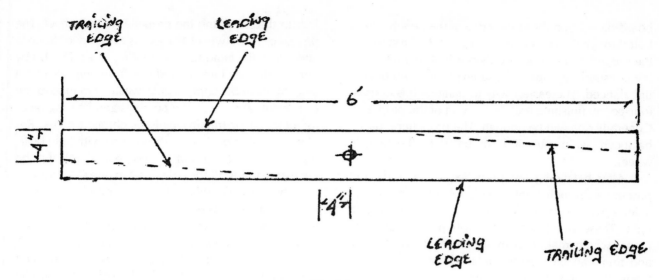

Fig. 6-5.

edge will be sharp to prevent turbulence behind the blade which will slow it down. Do the final shaping with a hand or power sander so no uneven dimensions are visible.

The propeller should be sealed with a good wood sealer and covered with at least two coats of varnish or outside house paint to protect it. When that is done, balance the propeller by mounting it on a

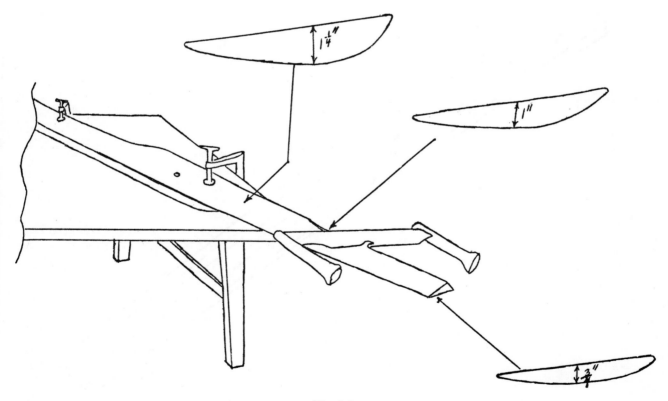

Fig. 6-6.

board nailed upright to the edge of the bench. Use a bolt through the center hole as a pivot. Make sure that the propeller turns free and then spin it several times. If it consistently stops in the same place, it is probably unbalanced. The easiest way to balance it is to drive four-penny finishing nails in the end of the propeller. Generally one will be enough. The propeller must be balanced again after it is mounted to the bicycle wheel.

After balancing the propeller, set it aside and procure a 26-inch front bicycle wheel with tire and axle. Also obtain a 12-volt bicycle generator and light. These small generators will put out about 2 amps, which makes them useful for 12-volt lights, and they can be used to trickle-charge 12-volt batteries. Very often these parts are available used. The generator and light kit cost about $12 new at this time while the wheel and tire may cost about $13 retail.

When the parts are procured, attach the wheel hub to the stand and the propeller to the wheel. The first step, if the wheel is used, is to remove the axle and check the bearings and bearing races. Do this by unthreading the locknut and hub from one side of the axle. Remove the components, wash them in gasoline or solvent, and check the bearings and bearing races for ʻchipped or missing balls or chipped races. Replace all failed parts, coat the bearing with light bearing grease, and reinstall the components. It will be necessary to install the axle offset to one side to adapt it to the stand. Do this by threading one bearing race on the axle as far as possible. (At the center of the axle is an unthreaded section that will stop the race.) Then reinstall the components in the proper order. Tighten the opposite race until the wheel is "set up" and then back it off slowly until the wheel revolves freely. The axle can be held in the hands to check this. Install and tighten the locknuts.

Next obtain a steel, heavy-duty $1/8 \times 2$-inch pipe

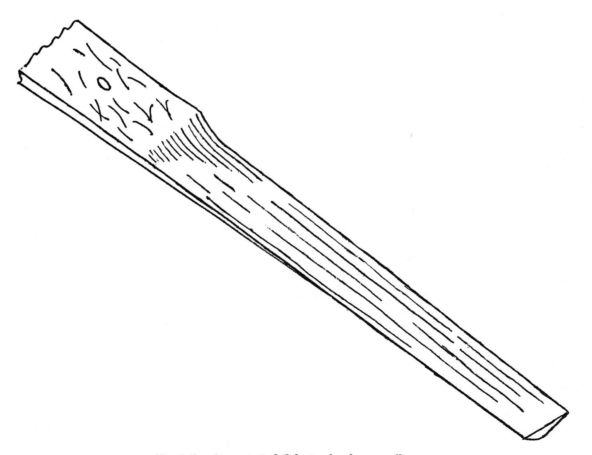

Fig. 6-7. Correct air foil design for the propeller.

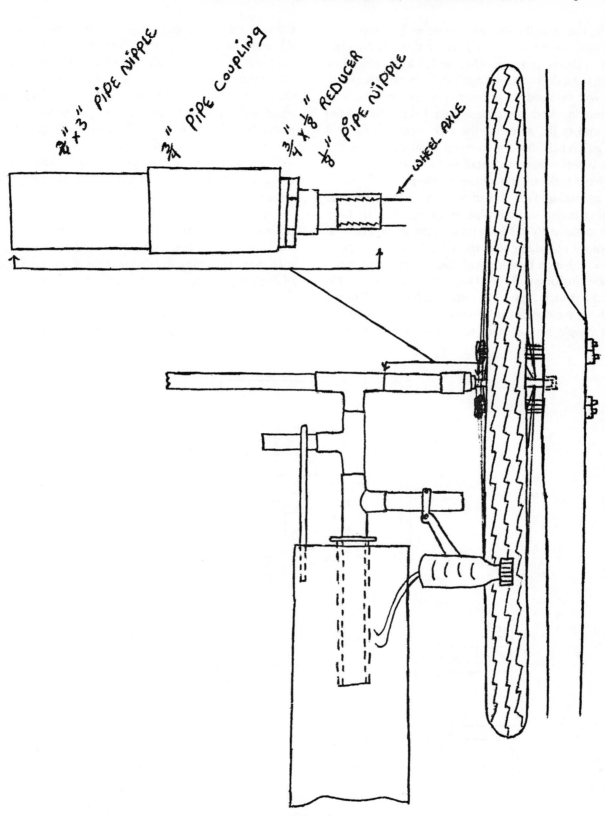

Fig. 6-8.

nipple, hacksaw the threads off one end, clamp it in a vise, and use a $5/16 \times 24$ tap to cut threads 2 inches deep in the sawed end. Then install a $5/16 \times 24$ hex nut on the axle and turn it against the bearing locknut. This will be used as a jam nut to prevent the axle unthreading from the pipe nipple. Finally thread the long end of the axle into the pipe nipple as far as it will go and tighten the jam nut against it. Now the cycle wheel is ready to install to the mount.

Next make up the mount for the propeller, vane, and generator. First thread a $3/4 \times 1/8$-inch reducer bushing into a $3/4$-inch pipe coupling (see Fig. 6-8). Then thread a $3/4 \times 3$-inch pipe nipple to the opposite end of the coupling. Next thread a $3/4$-inch pipe tee to the $3/4 \times 3$-inch bushing with the center outlet of the tee pointing down. Then at the center outlet of the tee thread a 3-inch pipe nipple and another tee. This tee will have the center outlet pointing to the rear. Next a close nipple and another tee with the center outlet towards the front, and finally a 6-inch pipe section threaded on one end only. Tighten all threads very well after the correct alignment is determined. If a joint tends to unthread after it is placed in operation, drill a .125 hole through the walls of both pipe sections and install a cotter pin. This is useful also for joints that can't be tightened in the correct alignment. When the mount is made up, thread the $1/8$-inch pipe nipple into the $3/4 \times 1/8$-inch reducer bushing. This will fasten the wheel to the mount. The next step is to fasten the propeller to the wheel.

Drill four $1/4$-inch holes through the hub of the propeller evenly spaced in a $1\frac{1}{2}$-inch radius around the center hole. Next drill a $3/4$-inch diameter hole 1 inch deep enlarging the center hole at the back side of the propeller to provide space for the end of the wheel axle.

Then select a sturdy 6-foot 4×4 hardwood if available. Find the center of one end and drill a $1\frac{1}{8}$-

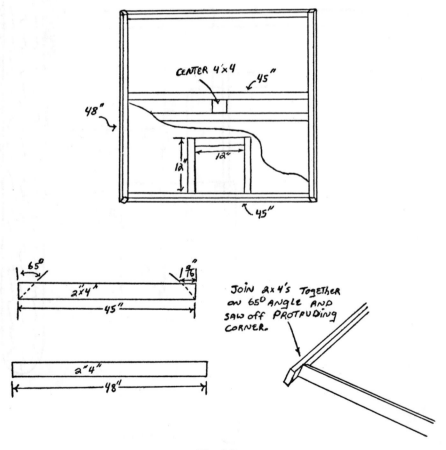

Fig. 6-9.

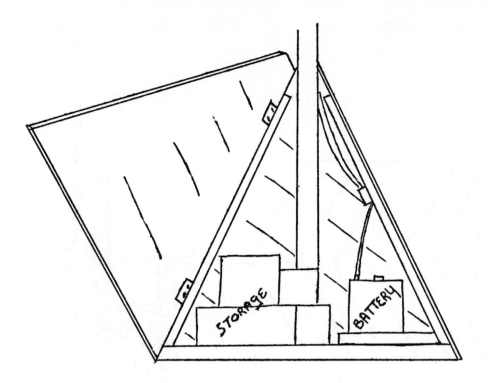

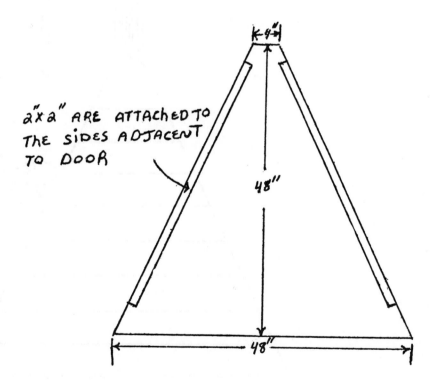

STORAGE

BATTERY

2"x2" ARE ATTACHED TO
THE SIDES ADJACENT
TO DOOR

48"

48"

Fig. 6-10.

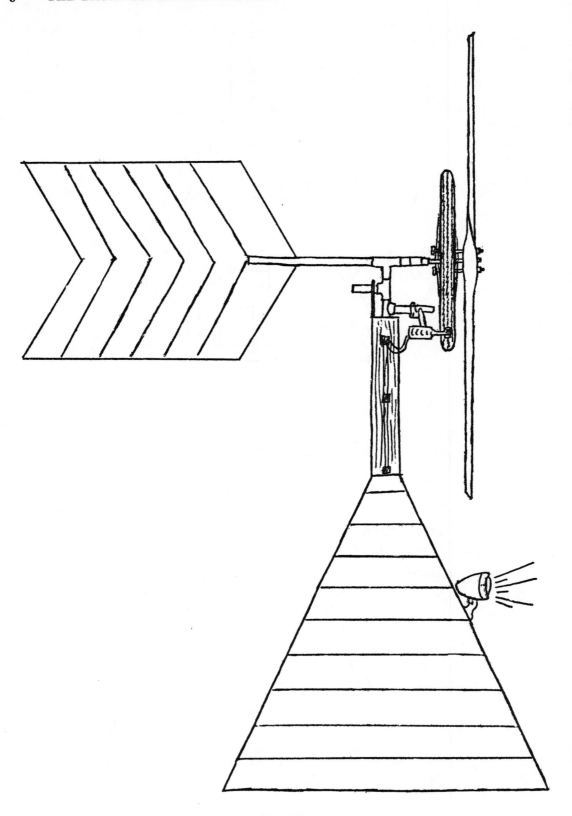

Fig. 6-11. Wind generator.

inch hole 6 inches deep directly on the centerline. Next cut 6 framing lumber 2 × 4's into the following dimensions: four 2 × 4's 45 inches long and two 2 × 4's 48 inches long. At each end of each 45-inch 2 × 4 cut a 1⁹/₁₆-inch triangle to form the correct 65-degree angle for the side walls of the stand (see Fig. 6-9). Spike two 45-inch 2 × 4's to the 4 × 4 flush with the end opposite the 1¹/₈-inch hole. Next nail the 48-inch 2 × 4's to the ends of the previously nailed 45-inch 2 × 4's. Finally slide the remaining 45-inch 2 × 4's into place and nail securely to form the square base. Saw off the projecting ends of the 48-inch 2 × 4's. When the base is complete, stand it up and nail in the floor.

The floor can be ¼-inch plywood or the equivalent. If a sheet of tempered masonite has to be purchased to make up the wind vane, the waste can be used for the flooring. When the floor is nailed in place, form a battery box centered at the surface which will eventually be the front. Next cover three sides of the base with scrap plywood. Considerable waste would result from using full sheets of plywood for this. The fourth side will be made from a sheet of 3 × 8 plywood, hinged and latched to form an access door. *Note:* if it is not desirable to charge batteries, with this unit no battery box need be installed. The space inside the base can be used for tools, toys, or gardening aids also. See Figure 6-10.

After the base is complete the propeller mount can be placed in the 1¹/₈-inch hole in the top of the 4 × 4. After considerable use this hole may become enlarged and allow the propeller to lean. If this happens drill out the 1¹/₈-inch hole to 1³/₈ inches and install a 6-inch length of 1¹/₈-inch steel tubing to use for a sleeve. After the sleeve is installed place the propeller and mount into place.

Since a propeller must face directly into the wind to be effective, a wind directional vane must be made up which will keep the unit facing into the wind (see

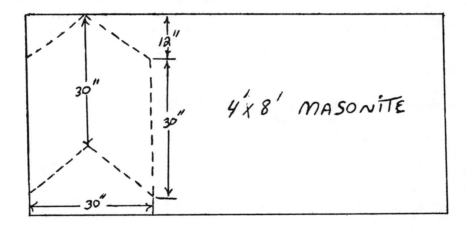

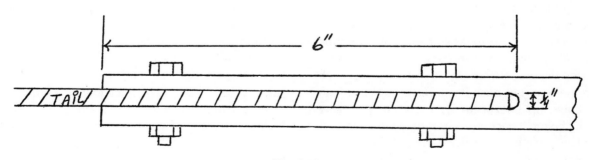

Fig. 6-12.

Fig. 6-11). Make the vane from ¼-inch tempered masonite. Streamline it as shown in Figure 6-12. Then procure a ¾ × 18-inch pipe section threaded on one end only. Measure off 6 inches from the unthreaded end and drill a ¼-inch hole directly through the center of the pipe. Next rotate the pipe 90 degrees and drill 2 ³/₁₆-inch holes directly through the center of the pipe. Finally, use a hacksaw to slot the pipe from the unthreaded end to the ¼-inch hole. When this is done the 18-inch pipe will have a ¼-inch slot 6 inches deep into the unthreaded end (see Fig. 6-12). Evenly spaced at right angles to this slot will be the ³/₁₆-inch holes. Next slide the masonite vane into the slot, drill through the ³/₁₆-inch holes, and install ³/₁₆-inch stove bolts to fasten the vane to the 18-inch pipe section. Then thread the pipe into the tee. Now the unit is ready to use except for installing and wiring in the generator.

Install a ¾ × 6-inch pipe nipple and attach the generator as shown in Figure 6-8. Remember that it is necessary to use two wires to connect the light to the generator for a complete circuit. Study the layout. If it is desirable to charge batteries with this unit, a "cutout" must be installed between the generator and battery to prevent the battery from discharging. Cutouts of this type are available from sources listed at the end of the chapter or build your own wind-activated device. No collector is needed for this wiring diagram since the unit is prevented from swiveling completely around, which would twist the wires, by stops drilled into the 4 × 4 base. Thus the wind charger should be set up facing the direction of the prevailing winds. Experience has shown that it will be inoperational only a tiny fraction of the time.

When the wind generator is done, move it to the windiest section of the garden or lawn and enjoy it. It is useful for illuminating a pool, lawn, or steps in the nighttime, as well as discouraging prowlers. During the daylight hours the smooth motion it produces will delight your eye each time the wind blows. Be sure to make up a simple hook to fasten the propeller during periods when windstorms are expected. If it is desirable to stop the propeller at any time, simply approach it from the rear and push it out of the wind. If high winds come up unexpectedly, it will simply blow over long before it becomes a hazard to life or limb. A simple fence can be constructed in front of the unit to keep small children from coming in contact with the whirling propeller.

WIND MOTOR

One of the greatest weaknesses of the propeller type wind motor is that it must be always facing the wind to be usable. Since the wind is constantly changing, this requires a mechanism to keep it oriented. Additionally, when high winds come up, the propeller should be stopped to prevent damage. The wind motor eliminates many of these weaknesses. It will run no matter where the wind is coming from and nothing short of a full-scale hurricane will do it any harm.

Moreover, it will work very well to turn a water pump or a propeller to keep a fish pond open in winter or produce a desirable current to cool it in summer. It is also useful for pumping water for a solar heating system or filling a container of a gravity-fed animal watering system. It also can be used as a source of free power for a workshop.

The following directions are for a wind motor using two halves of a barrel. For adding the two additional vanes when it is used for shop power just increase the length of the ½ × 3 center stock from 36 to 72 inches and add four additional strap iron braces.

MATERIALS LIST FOR WIND MOTOR

1. 55-gallon drum, either 1 or 2
2. 8 running feet of ½ × 1¼-inch strap iron and 24 ⁵/₁₆ × 1½-inch machine bolts, nuts, and washers.
3. 4 feet of 2½-inch pipe, 6 feet of 1-inch pipe
4. ¼ × 36-inch strap iron
5. Piece ¼ × 3 × 24-inch strap iron
6. Lengths of 1¼-inch diameter water pipes as needed to make the stand

The first step is to cut a 55-gallon steel drum in half. Use a cutting torch or a sharp cold chisel for this if you don't have a power saw. Make the cut from top to bottom. Then find enough strap iron to make four strips (eight strips for a shop wind motor) ½ × 1¼ × 24 (*B* in Fig. 6-13), and bend each strip at each end to form a 90-degree angle to fit the inside of the barrel. Drill through the strips and the barrel and bolt the strips to the barrel with ⁵/₁₆-inch machine bolts.

Next find a piece of metal stock ½ × 3 × 36 inches, 72 inches for a four-vane. Slot one end ¾ inch wide and 6 inches deep. Lay the other end on the strap iron strips (*B*) on one barrel and weld it in place. Then unbolt that strip from the barrel (to make

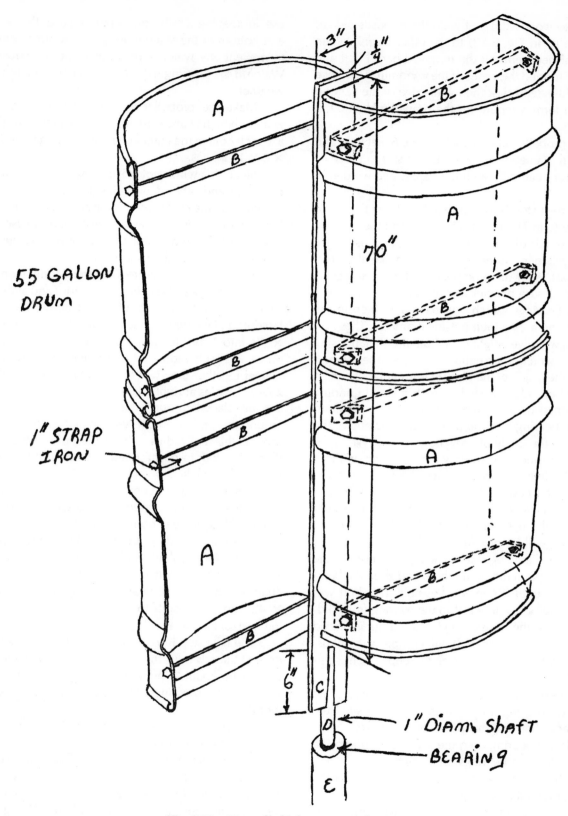

Fig. 6-13. Vanes for high-torque wind motor.

it easier to work with) and weld the opposite side of the piece (C in Fig. 6-13) to the other barrel. When this is finished, reinstall the removed barrel half and you have formed the vanes of the windmill. The piece (C) should extend past the lower edge of the vanes 6 inches to form a mount for the propeller shaft (D in Fig. 6-13).

The propeller shaft, which can be 6 feet long and ³/₄ inch in diameter, is placed in the slot in piece C, aligned very carefully with the vanes, and welded in place.

The next step is to secure a 4-foot length of 2¹/₂-inch pipe (E in Fig. 6-13). This will form the housing for the ³/₄-inch diameter shaft and will be used as a bearing retainer. Moreover, it will make the shaft freezeproof. After the pipe is secured, go to the bearing supply house or your favorite salvage dealer and find two bearings which will have a ³/₄-inch inside diameter and a 2⁹/₁₆-inch outside diameter. These dimensions will give a tight fit on both the shaft housing and shaft, which is essential for proper operation. If you find the inside of the pipe (E) is irregular, it can be smoothed or enlarged slightly with a small grinding wheel turned by a ¹/₄-inch drill. If it isn't possible to insert the bearing into the pipe, warm the end of the pipe with a fire made of a few newspapers. Well pipe expands considerably when warmed. Then, without getting the fingers burned, place the bearing in the pipe and let it cool and shrink to fit. If the bearing is slightly loose, it still can be used by drilling a hole completely through the pipe slightly below where the top bearing will be placed, but offset enough to miss the ³/₄-inch diameter shaft. Then put a bolt in the drilled hole, put a nut on the bolt, and slide the bearing in place until it contacts the bolt. This bolt will serve two purposes. It will stop the bearing from sliding down into the housing and when a nut is placed on the bolt and, tightened, it will compress the housing enough to grip the bearing to keep it from spinning.

The bearings have to be placed in this order. First put the top bearing on the shaft (D). Then slide the bearing and shaft into the top of the housing (E). Next, tap the lower bearing onto the shaft and up into the housing (E). If the bearings aren't tight, place a bolt through the housing and turn it up to compress the housing around the bearing.

If you are building the two-vane wind motor for use in keeping a fish pond open, coat both the top and bottom of the shaft housing with roofing cement to prevent any water from leaking into the housing. Water in the housing will cause it to freeze up in cold weather.

Make the propeller which will turn under the water when the vanes turn above the water by drilling a 1-inch hole in the center of a piece of metal stock ¹/₄ × 2 × 18 inches.

Since this wind motor is adaptable for turning a generator and other devices inside a shop, it can be mounted on the roof of a building and made very useful. Generally, it will be desirable to increase the torque of this wind motor by adding two additional vanes when it is used for shop work.

All other directions are the same. Mount it on the roof by cutting a suitable hole in the roof and slide the shaft down until it is about 1 foot above the roof. Hold it in place with a bracket made by sawing two 4 × 4's into 1-foot lengths. At the center of each, cut out a 2¹/₂-inch radius so that when the pieces are placed together they will form a 2¹/₂-inch bore. Drill 2¹/₂-inch holes at right angles to the shaft so that bolts can be placed through the 4 × 4 to tighten them around the shaft. Nail one section of the mount to the roof so that the radius adjoins the hole in the roof. Place the shaft in position and then bolt in the other half of the bracket to secure it. Paint the mount or cover it with roofing cement. Further, place liberal amounts of roofing cement around the areas of the roof which could leak from this installation.

Inside the building use a 2 × 4 frame to make a steady rest for the shaft. Nail the 2 × 4's to the rafters so they project down from the rafters about 4 feet apart. Connect the vertical 2 × 4's with horizontal 2 × 4's and use V bolts to clamp the shaft housing to the horizontal 2 × 4's.

The next step is to thread the end of the shaft if it doesn't already have threads. Use a ³/₄ × 16 die to cut threads on about 2 inches of the shaft. These threads will be used to connect the power shaft to the wind motor shaft.

The power shaft is a ³/₄-inch rod, long enough to reach from the end of the shaft to the floor. Rod of this type is available in almost any hardware store. Generally 4 feet will be needed. One end of this power rod has to be threaded also. The coupling between the rods can be easily made by drilling and

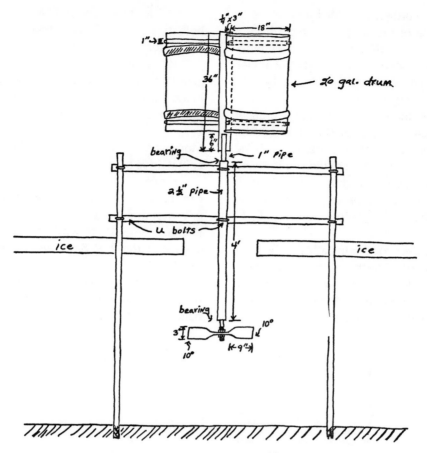

Fig. 6-14.   Wind motor used in fish pond.

tapping a 2-inch length of 1¹/₈-inch mild steel hex stock for the ³/₄ × 16 thread. If this is not desirable, make the connection by sliding a 4-inch section of ³/₄-inch inside diameter heavy-duty rubber hose over the ends of the shafts. Clamp the hose to the shafts with hose clamp as shown in Figure 6-15.

Now with the power shaft connected to the motor shaft, the next step is to make the lower bearing or steady rest which will keep the end of the shaft from "whipping." This lower bearing has to be made so it is easy to open for removing the shaft.

However, since it is low-speed, a hardwood bearing can be used. Make the bearing by drilling a 1-inch diameter hole in the center of a hardwood block measuring 12 × 6 × 6 inches. Further, drill 2¹/₂-inch holes at right angles to the 1-inch hole. Then carefully saw the block in half and fasten one-half to the floor with 8-inch lag screws so that half the 1-inch hole is in line with the bottom of the shaft. When the wind

motor is put in operation, the other half is bolted in place using ¹/₁₂ × 10-inch bolts, This will polish itself smooth and be a practically frictionless bearing with a little wear.

Now a pully/flywheel combination can be made up from an automobile tire rim. Generally, it is advisable to use the largest rim that can be found, since the larger the diameter of the pulley the more satisfactory it will be when driving a generator or alternator.

The first step in making this pulley is to cut off one rim flange. This will have to be done with a cutting torch, but very probably your friendly salvage dealer can do it for you when you buy the rim from him. Grind off any rough burrs left after the cutting operation.

Next make up the bracket for mounting the rim to the shaft. Do this by selecting an 8-inch section of ¹/₂ × 2-inch strap iron. Drill 2¹/₂-inch holes in the section 6 inches apart. Further, drill 2⁵/₁₆-inch holes ³/₄ inches on either side of the centerline. Next have a blacksmith heat and twist this section so a butterfly shape is produced with its end sections at right angles to the center. Acquire a 1½-inch-wide ⁵/₁₆ U bolt, nuts, and washers (see Fig. 6-15).

Mount the rim to the shaft by bolting the bracket to the rim. Then place the power shaft through the center of the rim and fasten it to the bracket with the U bolt. Be sure the cut-off section of the rim is facing up. If the wheel is off center, slightly slot a 1½-inch hole in the bracket and bend the bracket so that the rim can be moved to a true center position (see Fig. 6-15).

The rim can be used for driving a V belt by bolting another belt inside out to the center of the rim to form one side of a pulley. The bottom rim flange will then form the opposite surface. A more expensive way of driving a V belt with a rim is to bolt or weld a V pulley to the rim. Very little welding is necessary to hold this pulley; in fact, heavy-duty solder will be satisfactory.

When the pulleys are in place, the alternators or generators can be positioned and belts purchased or made up that will fit them. Remember direction of rotation can be changed with a belt by crossing it between the driver pulley and the driven pulley. Moreover, a vertical drive can be changed to a horizontal drive by twisting the belt.

Once the flywheel is in place, a rubbing block can be attached to the floor that will serve as a governor and brake. Slot a section of 2 × 8 as shown in Figure 6-15. Place washers over the bolt on the top surface of the 2 × 8 and use a wing nut to adjust it (see Fig. 6-15).

To have constant current available with a wind-driven generator, it is necessary to store the current in batteries. Batteries, of course, not only store current

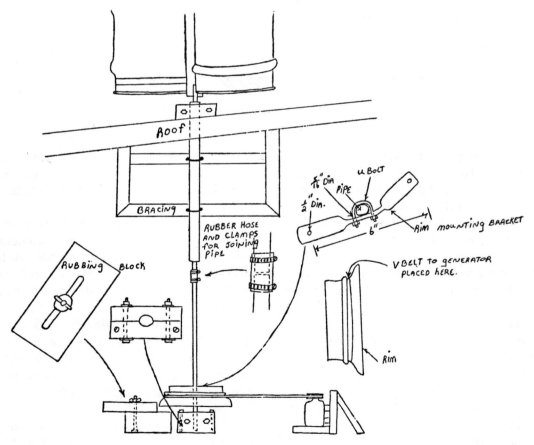

Fig. 6-15.   High-torque wind motor used in workshop.

but smooth it out, which allows appliances and tools to utilize it. The best batteries for this type of use are the nickle cadmium or airplane batteries. Nickle cadmium batteries are superior to lead acid batteries because they won't accept an overcharge and they remain efficient in cold weather. Generally, they last much longer also. However, they are much more expensive. If lead acid automotive batteries are used, a voltage regulator must be wired into the charging circuit to prevent overcharging the batteries and to prevent them from discharging when no current is being produced. Sometimes it's desirable to change the 12-volt DC current produced by an automotive-type alternator or generator to 110 volts AC. This can be done by wiring an inverter into the circuit. Inverters are widely used on campers and are sold by most mail order houses as well as many retail hardware outlets.

Electricity isn't the only energy that can be produced by this wind motor. It will function very well to turn the pump of an air compressor or to pump water. Also in hot weather the shaft can be fitted with a fan and it will help cool the shop. Many more applications for this wind motor are possible, such as turning grinders, lathes, drill presses, and small grain grinders.

Chapter 7

DOG HOUSES
AND PONY SHEDS

When I was a young lad it was a great delight to follow my dad around as he paid social visits to the dog raisers in the Wisconsin countryside. Some of these places were well kept up and the dog houses were clean and sturdy. Their peaked roofs were lined up like rows of soldiers, and not a thing was out of place. Something was amiss, though, since the dogs barked incessantly and nervously hauled at their chains.

Old Walt White raised dogs too, and his dog yard was littered and rather dirty. His dog houses were dingy boxlike affairs with flat roofs and they were scattered haphazardly around the yard. However, his dogs lay on the top of their houses in relaxed good humor and other than a few warning barks when a stranger approached they were quiet and contented. As I grew older and learned more about dogs, the reason for this became understandable. Each dog likes to feel that he is king of his castle. The single most important item in projecting that feeling of well-being is his house and in particular the

roof of his house. It should be flat and smooth so the dog can comfortably lie up there and feel as though he has command of the situation. This probably harks back to the predomesticated day of the dog when he would, like a fox or wolf, select a knoll over his burrow to nap on and watch the surrounding countryside for prey or foe.

It would seem that commercial dog house builders would have grasped this message long ago but apparently not. Just the other day I received a catalog from a large dog supply house and there was a new model dog house with a patented roof. Yup, you guessed it, the patented roof was as sharp as a spear. Obviously no one asks the dogs what they want.

ARCTIC DOG HOUSE

We kept this fact firmly in mind when we were building our arctic model dog house. The shed-style roof was made with only a 1-inch pitch from front to back. Further, we covered the top with #30 builder's felt and then installed 90-pound smooth roll roofing over that. This, of course, to provide a comfortable napping place for our beagle as well as hold the heat in and keep the rain out.

Also, to retain the body heat of the dog inside the house we made the roof like a sandwich with 1 inch of foam insulation between the panels. Foam insulation should be covered when used in a dog house since if it is exposed the dog will probably chew it. Besides destroying the insulation, the dog might swallow some of the pieces. Foam insulation is very hard to digest. The top also has to be made with hinges at the back so it can be raised to install straw bedding or to clean and disinfect the house. Finally, it should be light enough to be raised easily by anyone who might be pressed into service as a dog house cleaner.

Another important consideration in building a dog house is the size of the house in relation to the size of the animal. Not surprisingly the better a dog house "fits" the dog, the more comfortable it will be in cold weather. The house we designed was made to fit a 21-inch beagle and it should be satisfactory for all medium-sized dogs. For very large dogs, add 6 inches to the width and length. For very small dogs subtract 6 inches from the length and width.

MATERIALS LIST FOR ARCTIC DOG HOUSE

1. 2 8-foot 1 × 4's
2. 4 × 8-foot sheet of 3/8-inch exterior plywood
3. 37 × 46-inch section of 1-inch-thick foam
4. 12 square feet of #15 builder's felt, 12 square feet of 90-pound roll roofing
5. 2 8-foot 2 × 4's
6. 4 × 8-foot section of 5/8-inch plywood
7. 2 4 × 8 sheets of 1/4-inch exterior plywood
8. Approximately 40 running feet of 1 × 4 material
9. Insulation, nails, etc. as needed

Drafts and ground moisture have a comfort-destroying effect on most dog houses. To counteract this, our Arctic dog house is made with interior baffles and a swinging entrance door, hinged from the top, that can be opened by the dog as he enters. To counteract ground moisture the house is set on a foundation of treated 2 × 4's to provide a cushion of air between the bottom of the house and the ground.

Roof

Now for the actual construction of the top: First, obtain two 8-foot-long 1 × 4 boards and cut one 36-inch length and one 45-inch length from each. Nail the two longer boards across the ends of the shorter boards to form a rectangle. Nail them together with three eight-penny finishing nails in each end of each. Next, obtain a sheet of 3/8-inch exterior grade plywood and saw out a section measuring 39 inches by 48 inches. Square up the 1 × 4 frame and nail the piece of plywood to the frame with six-penny nails placed every 6 inches. Make sure the panel overlaps the front, back, and sides evenly (see Fig. 7-1).

Next, cut out and glue a piece of 1-inch foam insulation to the inside of the 1 × 4 frame. Make sure the foam fits well. Also use the flexible-style foam so it will form a tight seal on the sides of the house when it is in place. If the insulation doesn't fit the side walls well, a bead of putty should be placed around the top edges of the side walls before the roof is put in position. If foam insulation is not available, use a fiberglass batt, marsh hay, or cattail fuzz. This type of insulation should be protected with a 1/4-inch plywood panel or wire mesh. Nail the panel to furring strips placed around the inside of the 1 × 4-inch frame.

When this is done, the roof can be covered with a layer of builder's felt followed by a layer of 90-pound rolled roofing. Make sure that the edges of the felt overlap, the roofing is put on across the length of the dog house, and the edges are sealed with tar and well nailed. Also make sure any ends of the roofing are nailed down so the dog doesn't rip them loose.

When this is done, paint all the exposed surfaces of the top with a good wood preservative inside and out. Make sure it is nontoxic, however. The final step is to install a screen door hook on each side so the top can be fastened to the side of the house.

Floor

With the roof done, the floor can be the next item to construct. First make the frame. Select two 8-foot 2 × 4's and cut a 29-inch length and a 42-inch length from each. Drill two 1/2-inch holes 6 inches on either side of the center in each long 2 × 4 (see Fig. 7-1). Drill the holes at an angle so a direct movement of air is not possible under the house.

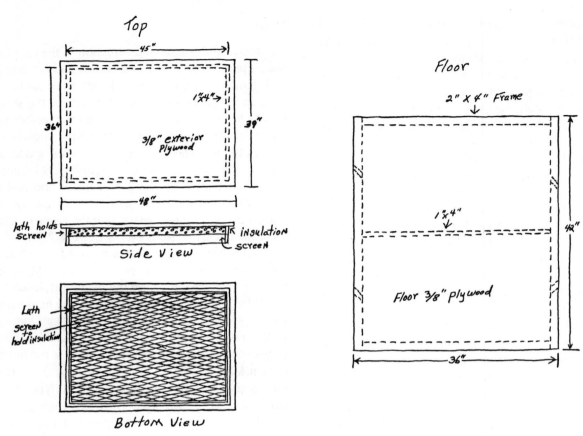

Fig. 7-1.

These holes are for ventilation under the house. Now nail the long 2 × 4's across the ends of the 36-inch 2 × 4's with two sixteen-penny nails evenly spaced in each board. Square this frame up and nail a 36-inch 1 × 4 board across the center of the frame. Make sure one ventilation hole is on each side of the center board. With the frame made, cut out a 36 × 42-inch piece of exterior grade plywood and nail it to the frame. Make sure the outer edges of the plywood are exactly even with the outside of the frame since the side walls overlap the bottom. Drive eight-penny finishing nails through the plywood into the 2 × 4's. Countersink the heads for a good fit. Some advantage can be gained by installing a vapor barrier under the floor so the cold air will be reflected away from the bottom of the floor instead of being conducted by it. This vapor barrier can be a single sheet of double-wall cardboard of the kind used to make boxes, covered on the bottom surface with aluminum foil.

The aluminum foil which comes in rolls for kitchen use is good for this. Just glue it to the cardboard with any good glue such as Pliobond. Be sure the cardboard fits the bottom very well and then cut the foil so it will slightly overlap the sides and the center crosspieces. Glue it well to the cardboard and to the sides.

The top surface of the floor should be sealed with a good wood preservative to keep moisture from soaking into it. Apply at least three coats. The floor then can be made almost immortal by covering it with a layer of floor and deck paint. Likewise, all other exposed surfaces should be well painted with a safe wood preservative.

Side Walls

The next step is to make the side walls. They are made double-walled with a 1 × 4 frame. The space between the sides of the frame can be filled with insulation (see Fig. 7-2), preferably foam insulation of the same type used to fill the roof. However, any kind of insulation can be used, including several panels of cardboard boxes stacked together. Make sure they fit tight, though. Make the frame for both sides at the same time. Start by cutting one 26-inch length, one 30-inch length, two 28-inch lengths, two 41-inch lengths, and two 41¼-inch lengths of 1 × 4 stock. Nail the frames together with two six-penny finishing nails placed in the ends of each.

Next, saw out the two outside panels of ¼-inch tempered hardboard or ¼-inch exterior plywood. These panels should measure 30 inches × 28 inches × 42 inches so they can overlap the bottom frame 2 inches and be nailed to it (see Fig. 7-2). Nail the panels to the frames and trim at the roof line. Install the insulation and nail the inside panels to the inside of the frame.

Ends

The ends are a little more complicated to make. First, to make the back end, nail a 1 × 4 to the inside of each corner of the sides (see Fig. 7-3). This should be inset ¼ inch from the outside corner. Nail it through the paneling into the 1 × 4 frame with eight-penny nails. Slant them slightly for maximum efficiency. Now the panel for the back can be nailed first on the outside and then on the inside. Note that the outside panel has to be 36 × 26 inches wide and the inside back panel is 29 × 26 inches (see Fig. 7-3). Insulate the back also. If vents are to be provided, they should be 1-inch holes drilled through the back wall; three is enough. Generally, they will not be needed. Insulate the back wall also. Now the front wall is the only one left. Nail a 1 × 4 to each front side wall so the outer edge of the 1 × 4 is flush with the outside edge of the piece. Next, cut out the inside and outside panels, clamp them together, and saw

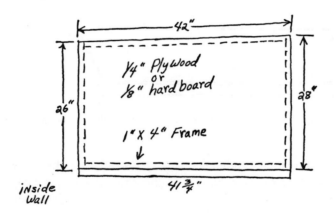

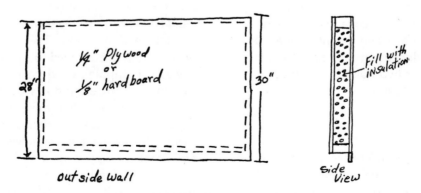

Fig. 7-2. Side wall of arctic dog house.

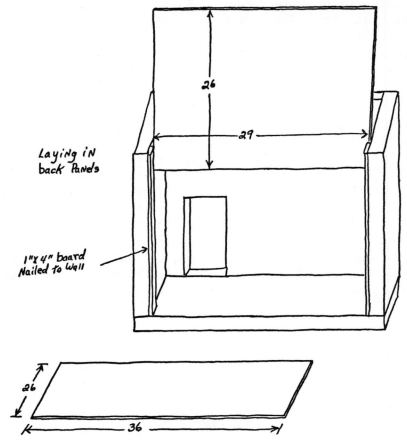

Laying in
back Panels

1"x 4" board
Nailed to wall

26

29

26

36

Fig. 7-3.   Laying in back panels in the arctic dog house.

out the entrance in both at the same time. Study Fig. 7-4. Unclamp the pieces and nail them in place. Now make a frame to line the entrance hole with, slide it into place, and nail through both the inside and outside panels to keep it in place. When this is done, fill the side walls with insulation. The tops of all the side walls are left open so no moisture will condense inside the walls. If you want to cover the insulation on the top to keep out insects and rodents, cut a 1 × 4 board to the correct dimensions and nail it in place. Be sure to drill one or two holes in it, however, to prevent moisture condensation.

Baffle and Door

The house is now complete except for installing the baffle and the door. Make the baffle (see Fig. 7-5) of ⅜-inch thickness plywood and metal tip the

exposed edges so the dog won't chew it. Next, cut out and install the door (see Fig. 7-6).

The door is designed to be opened by the dog with his paws and nose. Any dog should be able to learn to open it in a hurry. In fact, if your dog can't learn to open it by himself, perhaps you should suspect his basic intelligence. At first, however, it should be left open so the dog will be aware that it is the entrance to the house. Then lower it a little each day so he gets accustomed to opening it by himself. No weather stripping need be installed around the door entrance since the dog would probably tear it off, and even if he didn't, there is the danger that it would make the house too tight and not enough fresh air would be admitted. In warm weather this door can be removed or propped open.

Now, the dog has to be tied on a chain or this house has to be located inside a pen. Make sure that

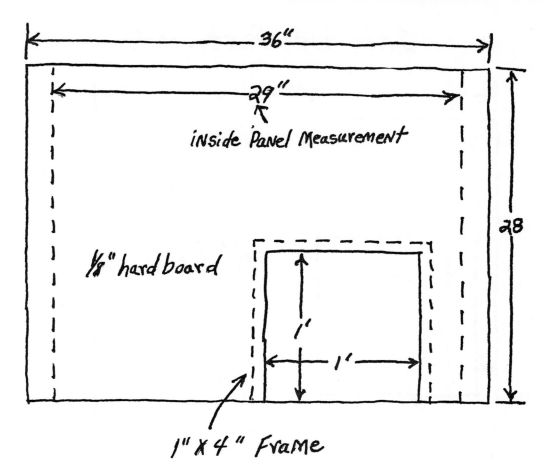

Fig. 7-4. Front panel of arctic dog house.

the chain is long enough since he can't, of course, sleep on the top if only the entrance hole is inside of his run.

Automatic Feeder and Waterer

When the rest of the house is done, the automatic feeder and waterer can be installed if it is desirable to have them. The advantages of an automatic feeder and waterer are obvious, of course. Overeating is usually not a problem. After the first week or so of gorging himself a dog will usually limit his daily intake of food to maintenance level. Automatic waterers are not practical in winter and a large pan of water should be kept available then. Be sure to stake your dog so he can just reach the house. If he is too close, he will tangle his chain around the house.

CHATEAU DE CANINE

This house will keep a dog comfortable in any weather. However, if you are going to use a wire run with the dog house outside the run or if you are looking for a dog house that will have considerable aesthetic value, the next model, which we call the Chateau de Canine, will be more to your liking. The high roof blends especially well into a wooded setting, and the space under the roof can be used for storing dog care items. No insulation is designed into this house, but it could be provided if desired by gluing foam sheets to the inside surfaces of the house. The baffle shown in Figure 7-8 is optional, but it will make a big house suitable for a small dog or it will prevent drafts from blowing onto the animal when he is in the house.

Interior Partition

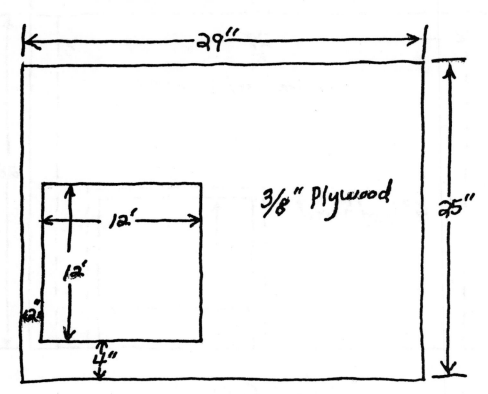

29"

3/8" Plywood

12'

12'

12'

25"

4"

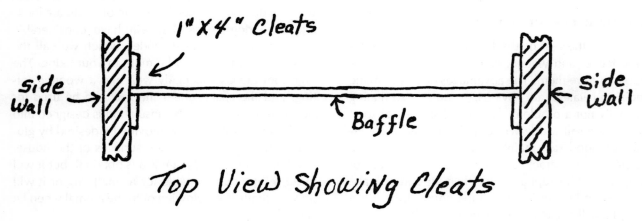

1" X 4" Cleats

Side Wall

Baffle

Side Wall

Top View Showing Cleats

Fig. 7-5.

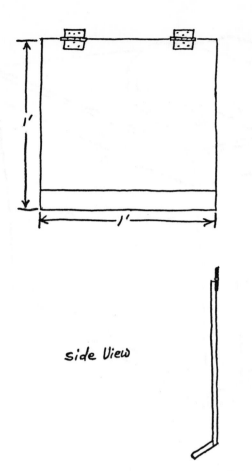

side View

Fig. 7-6. Arctic dog house door.

MATERIALS LIST FOR CHATEAU DE CANINE

1. 1 8-foot 2 × 4 board
2. 1 8-foot 2 × 2 board
3. 3 4 × 8-foot sheets ⅝-inch-thick exterior plywood
4. 1 × 4-inch furring strips as needed (8 feet)
5. 24 square feet asphalt or wooden shingles

Floor

The first step is to build the floor. The floor can be made of 1-inch lumber or ⅝-inch plywood placed over a frame. First, saw two 2 × 4's, 42 inches long. Chop the bottom corner on each end to make runners so the dog house can be slid along the ground. Next, cut three 2 × 2's to 33-inch lengths. Position one on each end of the 2 × 4's and one in the center (see Fig. 7-9). Nail through the 2 × 4's into the ends of the 2 × 2's with ten-penny nails, two in each joint. Make sure

the 2 × 2's are flush with the top surface of the 2 × 4's.

Next, saw out a panel of ⅝-inch exterior grade plywood 42 inches long and 36 inches wide. Position the panel over the frame and nail through the plywood into the frame with eight-penny common nails spaced every 6 inches. Be sure to nail into all parts of the frame. If 1-inch lumber is used, install it parallel to the runners.

When this is done, the floor is structured and it can be painted with a primer coat of nonlead paint or waterproof preservative. Be sure the preservative contains no toxic additives.

Side Walls

The next item to consider is the side walls. They also can be made from plywood. Simply cut two strips of plywood 9½ inches wide and 40⅜ inches long (see Fig. 7-10). A 1 × 10 board will also work. When the side walls are cut to the correct dimension, bevel one corner of each to 45 degrees. This bevel will be placed under the roof at the top edge of each side wall. When this is done, prime paint all raw surfaces of the side walls. Set the side walls aside while you make the ends.

Ends

The ends are identical except that the entrance hole is provided in the front end. Make the ends from plywood also. Proceed to lay out the ends on a sheet of plywood 96 inches by 48 inches by first laying the sheet crossways in front of you. Start at the lefthand bottom corner and measure up the side 10¾ inches. Make a pencil mark here (see Fig. 7-10). Now return to the starting point and measure along the bottom edge of the sheet to 35 inches. Make a pencil mark there. Next use a framing square to project a vertical line at least 10¾ inches long at right angles to the bottom of the sheet starting from the last pencil mark at 35 inches. Next, find the exact center (17½ inches) between the starting point (lefthand bottom corner) and the 35-inch mark and make a pencil mark there. Project a line from this last pencil mark on the bottom of the sheet of plywood to the top edge. This line will be 48 inches long. Use a straight edge to connect the last pencil mark to both 10¾-inch marks to finish out-

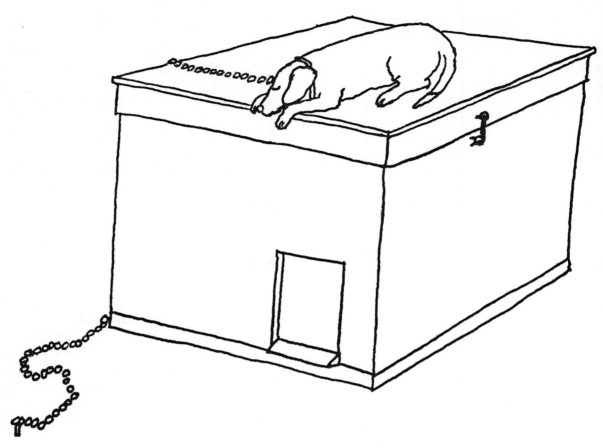

Fig. 7-7. Arctic dog house in use.

lining one end. Mark off the entrance hole also before sawing. Start at the centerline. Measure up 18 inches from the bottom edge on the centerline. Make a pencil mark. Next, make pencil marks along the bottom edge 6 inches either side of the centerline. Use a square to project a line 18 inches high at these locations also. Draw a light line to connect all the 18-inch marks to form a 12 × 18-inch rectangle. Next, find a compass, set it at a 6-inch radius, and scribe in a round top for the entrance hole (see Fig. 7-10). Darken all lines so they are easily visible and saw out the piece by first cutting out the entrance hole and then the complete end. For ease and accuracy in laying out the other end just flop the end you have already sawed out down on the sheet of plywood and mark around the outline. Then saw it out.

Baffle and Upstairs Floor

Now, cut out the baffle and the "upstairs" floor (see Fig. 7-11). The baffle is a sheet of plywood measuring 6 × 35 inches. It is nailed to the bottom of

the dog house to prevent drafts from creeping along the floor. The "upstairs" floor is 24 inches wide and 43 inches long. When it is nailed into position, the space between the roof and the upstairs floor will form a large storage area for the leash, manure scoop, worm medicine, flea powder, and other necessary items. Do not keep food in the storage space, however, as some dogs might wreck the house trying to get at it. To provide entry to this storage space, a door is cut into the end opposite the dog entrance. The piece cut out can be hinged. Use a turn button to keep it closed. Before the floor is nailed in, a corresponding piece of foam insulation can be glued to it if desired.

Roof

The roof panels are made from $5/8$-inch plywood or tongue-and-groove 1-inch boards. They measure 48 inches (see Fig. 7-12).

All projecting roof edges should be covered with

sheet metal to keep the dog from chewing them. Roof flashing, which is available in hardware stores and lumberyards, works well for this. Just cut it into 2-inch side strips, tap it into position, and nail it in place with screen brads. Although some tooth damage may occur to the center partition, it should not be covered with metal because it would be extremely cold in winter.

Assembly

When all of these parts are cut out, the house can be assembled. Start by nailing the ends to the floor with eight-penny finishing nails. Be sure they are positioned correctly. The bottom surface of the ends should be flush with the lower surface of the 2 × 2 crosspiece. Next, set the ends in place and toenail through the side panels into the frame of the floor. Further, nail the ends to the side panels. With this done, nail in the upstairs floor and then fasten the roof to the ends and side walls. If plywood is being joined to plywood, a good mastic should be applied first and the joint then secured with threaded flooring nails. If you do not desire to secure the roof this way, a very good alternate method is to nail furring strips to the

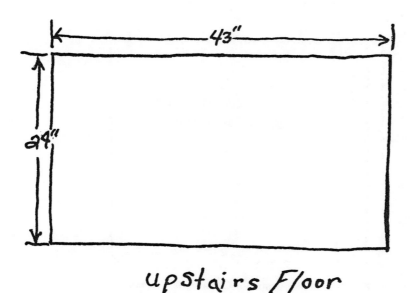

Baffle

upstairs Floor

Fig. 7-8.

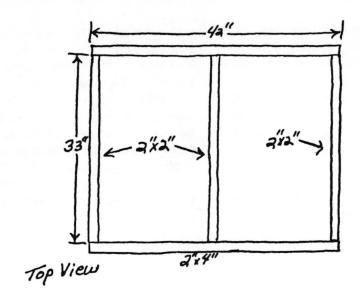

Top View

Side View

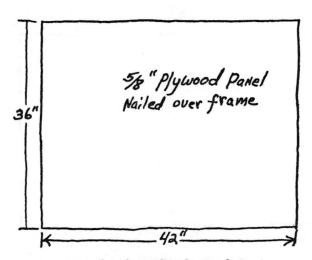

5/8" Plywood Panel
Nailed over frame

Fig. 7-9.   Floor frame for Chateau de Canine.

Side Panel

1 X 10 Board or
5/8" exterior Plywood
Both sides Typical

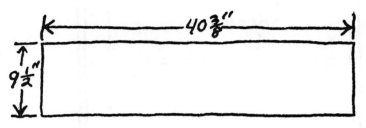

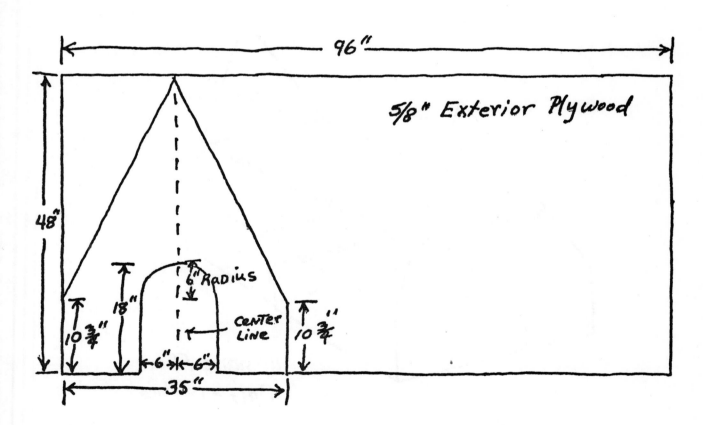

Fig. 7-10. *Top:* Side panel. *Bottom:* How to cut ends from plywood sheet.

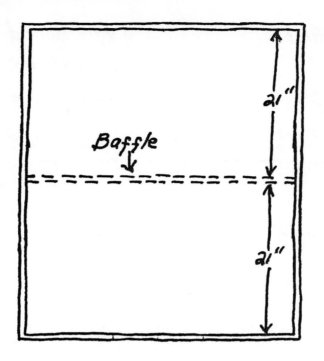

Baffle

21"

21"

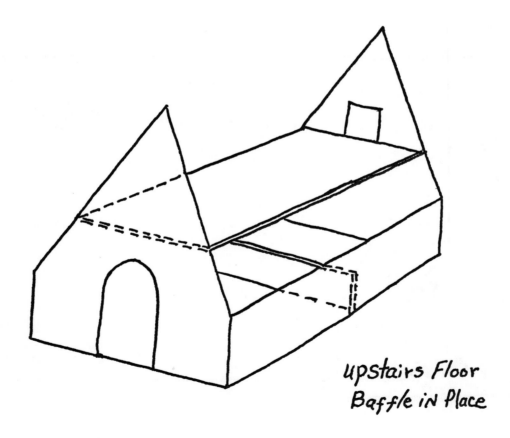

Upstairs Floor
Baffle in Place

Fig. 7-11.

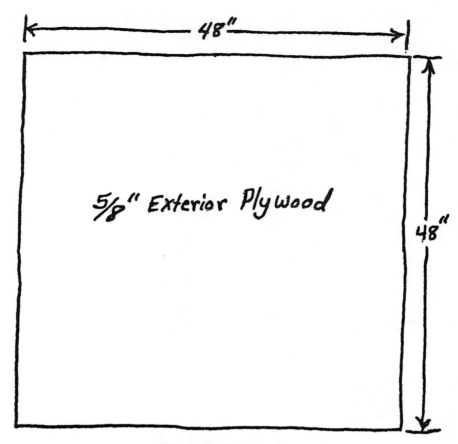

Fig. 7-12. Roof panel.

ends so they are flush with the edges. Then nail through the roof boards into the furring strips.

After the assembly is complete, the roofing can be applied. If at all possible, use wooden shake shingles for a truly rustic appearance. Consult chapter 5 for a method of splitting out your own shingles if you do not desire to buy them. Roll roofing of composition shingles will also work. Paint all surfaces cabin brown and trim the entrance, upstairs door, and roof edges in orange.

DOG PENS

With the shelter for the dog complete, some method of restraining him takes priority. A chain and collar are widely used, of course, but chains have limitations. They tend to get tangled up or twisted so that they keep a dog from moving freely. They also can be dangerous in lightning storms. Still another danger and a frequent cause of death among large

strong dogs tied with chains is that a wild animal or another dog will tease the tied dog so much he will forget he is tied and run full speed after the intruder until stopped suddenly by the chain. This can cause a broken neck or other injuries. Large active dogs should have a spring incorporated into their chains to cushion this shock. Heavy springs are available at hardware stores. They can be fastened into a chain with small cable clamps.

Another way of providing a shock absorber and also allowing the dog more freedom without an excessively long chain is to stretch a cable between two trees or posts and use a slip ring to fasten the dog's chain to the cable. Then he is free to run back and forth as far as the cable allows.

A better way of controlling a dog than any method of tying him is to build him a run so he is free to go anywhere within the confines of an enclosure. How big an enclosure should he have? Well, the ideal run would be to fence off your whole yard and

Fig. 7-13. Completed Chateau de Canine.

just let Ol' Rover rove. This, of course, is financially impractical. A more practical pen would measure about 4 feet wide and 20 feet long, which is adequate even for a large dog.

MATERIALS LIST FOR DOG PEN

1. 48 feet of poles for floor of dog run
2. 18 6-foot posts for support poles
3. 336 square feet of 12-gauge welded mesh wire
4. Bolts, nails, staples, door hinge and latch
5. Gravel or sand as needed for floor of pen
6. 120 square feet of roof sheathing and roofing

The sides of the pen should be at least 6 feet high and be made of chain link fence. Also the sides have to be inserted into the ground about 2 feet deep to keep the dog from digging out underneath them if the run floor is dirt. Concrete makes a better floor for ease of cleaning, but it is rather expensive and hard on the dog's feet.

Teetering on the ragged edge of financial disaster as we usually are here in the backwoods of northern Wisconsin, we learn to make use of natural material for most of our buildings. Accordingly, we built a pole pen for our dog run and we were delighted with it when it was done. It should fit well in any rural or semirural setting, it's very economical to build, and, most importantly, it keeps the dog secure and contented. The following paragraphs tell how we did it.

Staking Out the Enclosure

First, we staked out an area 4 × 20 feet over a deposit of sand, which would provide good drainage. If we didn't have sand, we would have had to dig out the soil down to about 2 feet and fill the excavation with coarse-grained sand or gravel. Even so, we excavated the soil to a depth of 12 inches in the pen, placed 6-inch poles along the perimeter of the excavation (see Fig. 7-14), and filled in the rest of the hole with washed gravel from the streambed. The poles were staked in place with 1 × 4 stakes driven into the ground before the gravel was deposited. The stakes were nailed to the poles. Care was taken that the outboard surfaces of the top row poles were kept even so the bottom of the pen wire could be stapled to them. The purpose of the poles, of course, is to keep the dog from digging under the wire at the sides and ends. If no poles are available, 2 × 12's could be used. This completes the floor of the run.

Building the Frames

The next step was to build the frames for the pen. First we cut three 4-inch posts 6 feet long (2 × 4's will work also) and squared one end of each. Two of the posts were left intact, and the third was used to make two 2½-foot pieces and one 1-foot piece. All posts

were painted with a commercial creosoted Penta waterproof preservative. Next, to assemble the frame, we found a level section of ground and laid out the lumber pieces with the 6-foot posts parallel to each other 4 feet apart. Next we laid out the 2½-foot pieces under the 6-foot and then laid the 1-foot between them so each joint overlapped (see Fig. 7-14). The purpose of this action was to mark for the notches. When the notches were marked, they were cut halfway through each log and unwanted wood was chipped out with a wood chisel. Next, the logs were placed in the assembly position again and 5/16-inch holes were drilled through each joint, a ¼ × 8-inch flathead bolt was placed through each hole, and a nut and washer were installed to fasten the joints together. The nuts were tightened just enough to make the joints rigid.

Place the frames in holes 2 feet deep on each side of the poles at the edge of the run. First, stake out

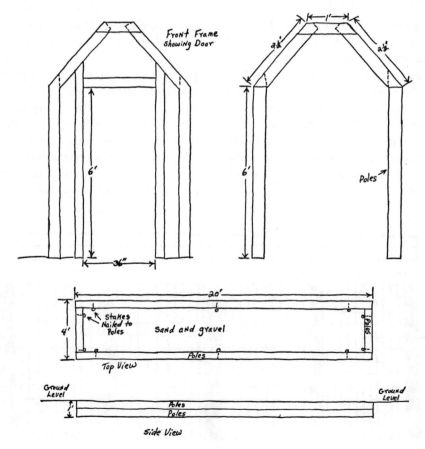

Fig. 7-14. Dog pen.

the posthole locations by carefully measuring each side and by placing a stake every 4 feet. Then use a posthole digger to dig holes just slightly larger than the corresponding posts at each location. When each frame is done, set it in the holes and backfill the holes just enough to hold the poles rigid. Then drive a nail exactly in the center of the 1-foot piece at the top of the frame. Hang a plumb bob from this frame and adjust the frame so the plumb bob falls directly in the center of the run. Further, use a level to make sure the posts are perfectly vertical. Then finish backfilling around each post and tamp them very well. After the frame is adjusted and placed correctly, tighten the ¼-inch bolts until they start to compress the wood; then drive three nails into each joint to further stiffen it. When this is done, paint the frame to match your color scheme. Repeat this procedure for each of the five remaining frames.

Wire, Roofing, and Screening

When all the frames are in place, install the wire, which should be 2 × 4-inch mesh, 12-gauge welded mesh, or the equivalent. The bottom strips should be 48 inches wide to avoid joints. Be sure to staple the wire to the poles at the edge of the run. The top of this type pen is commonly covered with boards and roofing paper to form an all-weather exercise area. Also in summer the sides of the pen can be screened with inexpensive fiberglass screen fastened to the outside of the wire to keep disease-carrying mosquitoes and flies away from the animals.

There are arguments on both sides of the question of putting the dog house inside or outside the pen. Inside the pen the house will take up run space and be harder to clean and bed. Outside the pen the dog can't sleep on the roof and there is some danger of his being stolen or let out through the hinged roof. If the house is placed outside the pen, place it at the end opposite the door, staple the wire to the house, and then cut it out for an entrance hole.

Door

The door of the pen should cover the full width of one end of the pen. An outside garage door can be used for this, or you can build your own according to the following method. Build a frame of 2 × 4's or 4-inch poles. Brace it with a diagonal brace from the lefthand to the righthand side. Cover the frame with wire mesh or make it solid with plywood panels. Install two large T hinges located 6 inches from the top and bottom. Install a hasp lock and keep the door padlocked except at feeding time.

All sorts of automatic dog feeders and waterers are available from the various dog supply catalogs or your nearest pet shop; at the least, an automatic waterer is recommended. Don't neglect to talk to your pet and pat him every day. His whole world consists of watching for you and trying to be near you. Don't disappoint him by neglecting him day after day even though you may be busy.

There are millions of dogs in this country, and each and every one deserves a good home. However,

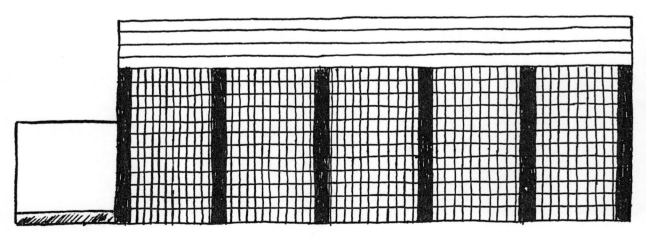

Fig. 7-15.   Dog house and pen.

many dog owners have other livestock and, especially in the suburbs and semirural locations where the household boasts one or more children, the pony is a favorite.

PONY SHED

A pony is, of course, a small breed of horse. A true pony must be less than 57 inches in height. Probably the best known of the different types of ponies are the Shetlands. The Shetland ponies, originally brought to the United States from the Shetland Islands, are rugged, patient, docile animals, very well suited to being ridden by children. Generally, when a child thinks of a pony, he thinks of a Shetland. The original Shetlands were kept out-of-doors most of the time. It is true they will survive that way, but they will get thin and look rough and ragged, and they may also get mean.

Far better than letting them range outside is to build a stable for them, especially since the cost is low and the building can be small. A building with outside measurements of 10 × 12 feet is sufficiently large for a single pony. This size building will provide a box stall for the pony, separate space for his saddle and other gear, and a storage space for some of his food. If hay is to be kept available on the premises, a separate hay shed connecting with the building can be built. Ponies love hay and will eat it almost continuously if it is kept in front of them. For this reason it is not advisable to use hay for bedding since the pony will eat up his bedding even if it is dirty. Wood chips, peat moss, peanut shell fibers, or sugar cane fiber make better bedding. Oat or wheat straw also is better than hay but ponies will feed on straw to some extent.

While we are on the subject of feeding a pony, it might be well to mention that a pony, like a big horse, will do best on timothy hay. A pony will consume about a forkful of hay a day; this should be fed on the floor of their stall or from a rack. Oats, of course, is still the best grain for a pony. Ponies should have about four quarts a day when they are being used or ridden every day. Children tend to feed ponies apples and carrots as well as sugar lumps. Ponies should not be fed too much sugar. However, salt is most important, and they should have a block of salt kept in their manger or a salt block in their pasture. Make

sure the pony cannot reach the grain bin since he will eat so much he will make himself sick or even kill himself—founder himself as the "horsey" set calls it. If possible, provide pasture since a pasture is a wonderful aid to a pony's well-being and general livelihood. About 2 acres of pasture is just right for a pony. Almost any kind of fence, except a single-strand electric fence is fine for ponies as long as it is 4 feet high. More about that later.

Foundation

Start building the pony shed by leveling the ground as for any building. Stake it out, find the corners, and place a cement block foundation around it. See chapter 2 for the proper way to erect the foundation. Be sure the leveled-off place measures 12 × 12 and the building sets straight with the street, road, or other buildings. The foundation blocks should be below the hard frost area, which is about 2 feet deep in cold climates. The location should be naturally well drained or artificially drained with tile placed underground. Failing this, the area should be landscaped to promote drainage of all standing water. (This stable is large enough also for use as a foaling stall.)

MATERIALS LIST FOR PONY SHED

1. Approximately 112 8 × 8 × 16-inch concrete blocks
2. 12 ³/₈ × 6-inch carriage bolts and anchor bolts
3. 27 2 × 4's for studding
4. 10 16-foot 2 × 4's for rafters
5. 12 sheets of ³/₈-inch exterior plywood or the equivalent for wall sheathing
6. 200 square feet of ¹/₂-inch exterior plywood or the equivalent for roof sheathing. 200 square feet of roof cover such as asphalt shingles.
7. Dutch door, tack door, 2 feed room doors, window. Hardware for the doors.
8. 4 pounds ten-penny common nails, 1 pound sixteen-penny common nails, 4 pounds ¹/₂-inch roofing nails

If it is not desirable to use cement blocks, a very economical foundation or footing can be made by digging a trench, filling it with stones, and pouring enough concrete around the stones so they weld together in a solid unit. Make sure the top is level

Back Frame

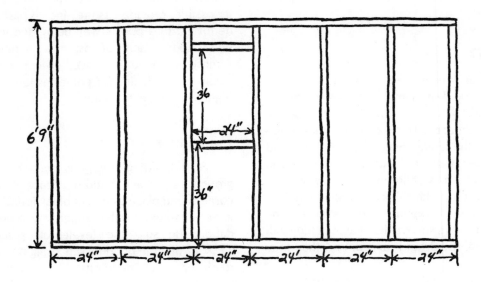

Front Frame

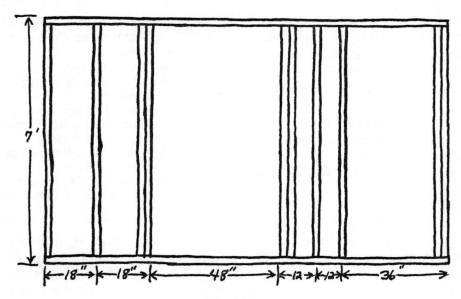

Fig. 7-16.

enough to take the plate, though. This, of course, doesn't mean it has to be completely level since concrete can be flushed underneath the framing shoe to level it. Be sure to place anchor bolts in the concrete so the shoe can be bolted down.

Framing

The framing is not standard; it has 2 × 4 studdings on 24-inch centers. The plate and shoe need not be doubled except at the builder's discretion. A good way to do the framing is to complete all the frames and then hoist them into position and nail them together. Start by building the back frame (see Fig. 7-16).

First, select two 12-foot 2 × 4's for a plate and a shoe, and square the ends so that each piece is exactly 12 feet long. This is necessary since yard run lumber is often too long and is not square at the ends.

Lay them out on a level surface close to the building site and separate the pieces by about 7 feet.

Next, select seven 8-foot 2 × 4's and cut each to 6 feet 9 inches. Be sure to square both ends. Now nail the plate and sill across the ends of the 6-foot, 9-inch studdings. Use two ten-penny nails for each end of each stud. The window frame is made by nailing 2 × 4's across between the studdings to create a rough opening measuring 24 × 36 inches (see Fig. 7-16). This completes the back frame.

The front frame is slightly more complicated. First, cut two 10-foot 2 × 4's for the shoe and plate as before; then cut ten 7-foot 2 × 4's. Nail one 2 × 4 at each end of the shoe and plate. Then measure 36 inches from the lefthand outside stud and install a double 2 × 4 studding. Further, measure off a 48-inch space and install a second double stud. The remaining two studs are evenly spaced between the double studdings and the outside frame.

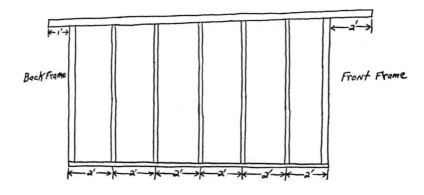

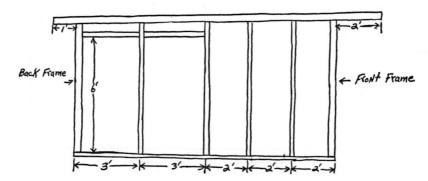

Fig. 7-17. *Top:* **Righthand frame.** *Bottom:* **Lefthand frame.**

The side frames are made thusly. First, select two 16-foot 2 × 4's. Cut each to 15 feet to make the two outside rafters. Next, prop the front and rear frames up and nail them together with the 15-foot 2 × 4's positioned so there is a 2-foot overhang in front and a 1-foot overhang in the rear (see Fig. 7-17). Nail them to the front and rear frame with rafter nails placed through the roof rafters. If desired, all the rafters can be installed at this time.

Roof rafters should be placed on edge every 16 inches apart across the roof. Next, set up a 2 × 4 studding in the left side wall to form a 6-foot-wide doorway. Do this by bolting a shoe to the foundation. Then nail a double stud between the shoe and the outside roof rafters. Further, space studdings on 2-foot centers to complete the wall. A single stud is used in the center of the 6-foot opening for a door stop.

The lefthand side frame is a complete frame with the 2 × 4 studs on 2-foot centers.

Sheathing, Roofing, and Insulation

The sheathing can be ⅜-inch plywood or 1-inch boards. If no siding is used to cover the sheathing, paint it with at least two coats of a good outside paint. Cover the roof with 90-pound roll roofing or the equivalent. If the climate is extremely cold, this building can be finished inside and insulated between the studs to create a warmer building. The doors will probably have to be handmade since purchasing doors of this style is doubtful.

Dutch Doors

Building dutch doors is more fun than work. First, procure a piece of 4 × 8-foot plywood, ¾ inch thick. Saw it in half. Nail a 1 × 4 across it so that it overlaps the bottom section by 2 inches. Further, nail

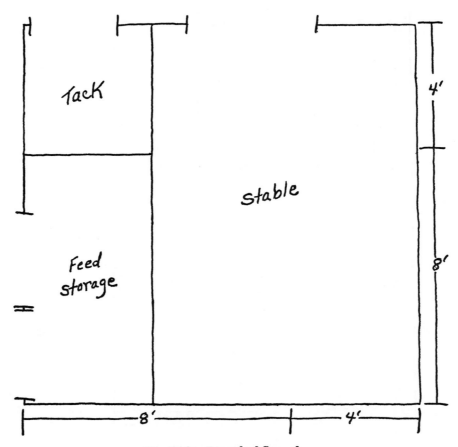

Fig. 7-18. Pony shed floor plan.

Fig. 7-19. Pony shed.

1 × 4's to form a crossbuck. If desired, the door can be covered with 1-inch-dimension lumber also up and down to form an up-and-down pattern the same as the siding. If you are going to stable an especially aggressive pony or a kicker, the door can be made of nominal 2-inch stock.

Floor

The floor in the stable should be left as packed earth if possible since standing on concrete is hard on horses' hooves. If you do desire a concrete floor, cover it with planks for the animal to stand on and slant the floor enough so the urine will drain away.

Feed Bin

A bin can be built in the feed storage room to keep oats in, and of course, several bales of hay can be kept there also.

Water and Heat

If desired, a cold water line can be run in underground and an automatic waterer can be furnished for the pony. Also an electric heater, such as a milk house heater, can be hung overhead for keeping the mare and foal warm when the mare foals. On very cold days some supplementary heat might be needed by any pony.

If you desire to feed oats to the pony from the feed room, a hinged door can be made to accomplish this. However, make sure it is horseproof because if the pony can figure out how to open it he might eat so many oats he will kill himself. Generally, it is best to leave the pony loose in this box stall, but of course, if he must be tethered, be sure the rope is short enough so he can't get his hoof over it. Many horses are tethered with a short ring to a cross rope so that they can have the freedom to move back and forth quite a lot without the danger of a long tie rope.

Exercise Yard

If at all possible, a large exercise yard should be provided for the pony, and a pasture of at least 2 acres is desirable. Naturally this should be fenced off, and the most desirable way to do this is with a wooden horse fence. This should be 4 feet high. Don't use barbed wire as the pony may cut himself badly on it in the spring when he is rubbing his long hair away. By the same token, be sure the fence has stout posts so the pony won't push them over when he starts rubbing in the spring.

Chapter 8

PLAY BUILDINGS FOR KIDS

Build any one of the projects contained in this chapter and the kids will no longer have to leave home to find a place to play. In fact, you may even "gain" a child or two. By all means let the kids help build these projects since they will learn important skills, develop a protective attitude towards the finished project, and derive more satisfaction from playing with it. An attempt has been made to provide information for building something for a child whether he has a bias towards rivers, the frontier, or outer space.

KENTUCK KAMP FORT

The Kentuck Kamp fort is built along the lines of an Adirondack shelter. From a practical standpoint it fits in with many lawn schemes wherever rough-sawn boards or large natural stones, logs, or ties are used in any building or decoration. If you have a wooded set-

ting available as a location for this building, by all means build it there. It will fit in so well that your boys might not ever come home again—except for meals, of course.

The Kentuck Kamp was originally used by the fast-traveling woodsman Daniel Boone a long time before anyone ever heard of the resort area commonly called the Adirondack Mountains today. Nor was this camp an original design of Boone. He reportedly was taught to build it by the North Carolina supply wagon drivers that the English General Braddock employed during the French and Indian wars.

Daniel soon grew adept at building one, however, and he "throwed" one up whenever he was going to stop for a few days. This delighted the settlers who followed Boone since they watched for and gratefully utilized the shelters that he left as he moved on. It is also rumored that he once said bitterly that if he had had time to build a Kentuck Kamp, the hostile Indians would never have been able to seize his son James, whom they tortured to death.

Several years later when Boone went to settle the Kentucky Territory, he lived first in a Kentuck Kamp and only reluctantly moved to a large house when he became prosperous. In a few years Boone was back in the Kentuck Kamp at Point Pleasant, West Virginia, however, since conniving lawyers arranged to have him stripped of his huge land holdings in the Kentucky Territory. It is one of the great tragedies of history that this brave man who settled and opened so much hostile land to settlers should have lived in poverty most of his life.

The original design of the Kentuck Kamp had the front wall left out. This made it possible to heat and light the interior of the structure with an open fire built in front of it. It was not exactly foolproof that way, however, since wayward winds occasionally swirled smoke and sparks onto the bedrolls of sleeping campers. Also, invariably some hardy camper would have to rise up during the night and replenish the fire. We included the front wall in our design for obvious reasons.

MATERIALS LIST FOR KENTUCK KAMP FORT

1. 24 8 × 8 × 16-inch concrete blocks or approximately 10 cubic feet of cement
2. 12 ¼ × 8-inch carriage bolts
3. 4 8-foot 2 × 6's for floor stringers
4. 100 running feet of 2 × 4 for floor joists
5. 3 4 × 8 sheets ⅝-inch plywood or the equivalent for flooring
6. 14 8-foot and 2 6-foot 2 × 4's for studding. 4 12-foot 2 × 4's for plates and sills
7. 7 10-foot 2 × 4's for rafters
8. 3 6-foot 2 × 4's for overhang
9. 250 square feet of half-log siding
10. 150 square feet of roof sheathing and asphalt shingles or roll roofing
11. Approximately 3 pounds ten-penny common nails. Door
12. Inside furnishing, etc

Site Selection

Some special considerations are apparent when selecting a site for the Kentuck Kamp playhouse. The most important is to avoid building beside a driveway, highway, or railroad track. Deep water, dangerous animals, and people should also be taken into consideration when selecting a site. Some other common but often overlooked situations to avoid are areas where water stands after a rain or where the yard will be muddy as a result of underground moisture. Also, do not build too near large dead trees or electric lines. Green trees, however, are desirable since they provide shade and a natural setting. As in all building, if your township is zoned, it is wise to consult your building inspector even before you start.

Foundation

The cheapest way to make a foundation for such a structure is to build a piling at each corner, although Boone probably just set his on stones. When you want to build a piling for a minimal-weight building such as this, it is not necessary to sink the piling in the ground more than 1 foot or to mineral soil. If the soil is too light or moist, the piling can be made so that it is flared on the bottom for maximum flotation. Anyway, start the form for the piling by digging an 8-inch diameter hole. Further, form a cylinder of heavy roofing paper that will just fit in the hole and project 6 inches above it. This is the form for the concrete. A tin can with both ends cut out will also work for a form.

Use the 1-2-3 mixture for pilings of this type. This mixture is one part portland cement, two and one-fourth parts sand, and three parts small gravel. Small amounts of concrete such as this can be mixed in a pail or even on a wide board. Mix the ingredients dry until they are well blended; then just add water until the mixture is pliable. Be sure to tamp it in the form so the air bubbles are removed from it. After the mixture starts to harden, set one ¼ × 8-inch carriage bolt in the center of each outside piling. The bolt head should be sunk about 3 inches in the concrete.

Be sure to level each piling again after the bolts are inserted, and do not attempt to build until the concrete has cured for three days. The concrete should be kept covered for at least twenty-four hours to allow slow curing. Concrete blocks can also be used. Place one 8-inch block underground and top it with another. Fill the center core with concrete and set the bolt in it.

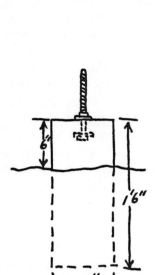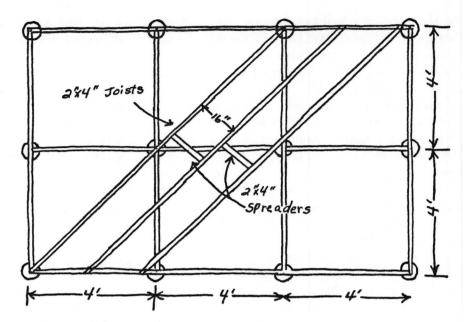

Fig. 8-1. *Left:* Piling. *Right:* Girders and floor joists.

Floor

After the pilings are set, build the floor frame. The first wooden members are 2 × 6 stringers placed on edge from one piling to the other. They have to be drilled for the 1/4-inch carriage bolts. Use a flat washer under the nut on the bolts and tighten the nut until it just starts to compress the wood. The floor joists are sturdy knotfree 2 × 4's nailed on 16-inch centers diagonally between the stringers. The two center joists in each set have spreaders placed between the longest joists.

The flooring is 5/8-inch plywood or composition board. Of course, 1-inch boards can also be used. Be sure to use well-seasoned boards for the floor and treat them with a good preservative prior to installing them.

The shoe and plate are single 2 × 4 stock and all studding is 2 × 4 also. The frame for the overhang is likewise 2 × 4 stock. The waste from the rear wall studding can be utilized in the overhang.

Rifle Ports

All windows are made like rifle ports. They are 12-inch diameter holes sawed in the side wall sheathing. A cover is made for each porthole so that it can be pivoted down to cover the opening when the building is "under siege." The cover pivots on a 1/4-inch carriage bolt. This bolt has the head outside and the threads inside the building, where it is equipped with a wing nut so the covers can be dropped from the inside. Each wall has four portholes. These portholes can be covered with screen if the danger of "attack" isn't too imminent.

Wall Frames and Rafters

Build the front wall frame by sawing out a plate and shoe 12 feet long. Nail 6-foot 9-inch studding on 24-inch centers between the shoe and plate. No allowance is made for the door since it is hung between the two center studs. See Figure 8-2. The rear wall of the Kentuck Kamp is made the same way except that the studs are 3 feet 9 inches long. See Figure 8-3.

The simplest way to assemble the frame is to make the front and rear wall frames first, then prop them up and lay each outside rafter in place. Note that the rafters have to be notched for the front and rear wall plates (see Fig. 8-4). When the rafters are nailed in place, spike the front and rear frame shoes to the floor and then finish installing all of the rafters. The rafters are 9 feet 6 inches long and have 12

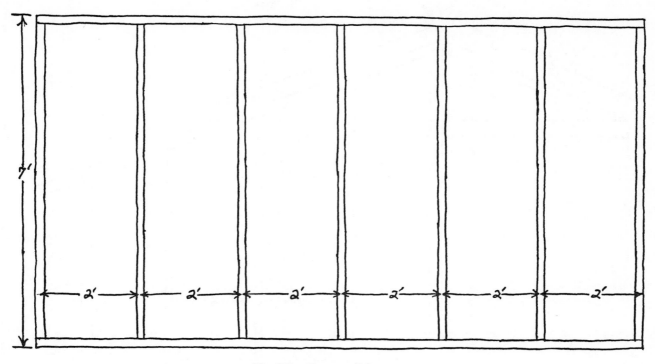

Fig. 8-2. Front wall frame.

inches of overhang at the rear. Once they are all placed, the side wall studs can be set in place, plumbed, marked, and sawed. The side wall studs are also placed on 2-foot centers. No stud is needed at the front and rear and no shoe or plate is needed on the side walls since the studs are toenailed directly to the floor and rafter.

Siding

No sheathing need be used with the Kentuck Kamp, instead, the siding is nailed directly on the studding. Half-log siding is available from most lumberyards. Sawmill slabs can be peeled and used also, and of course, rough-sawn boards directly from the sawmills can be used.

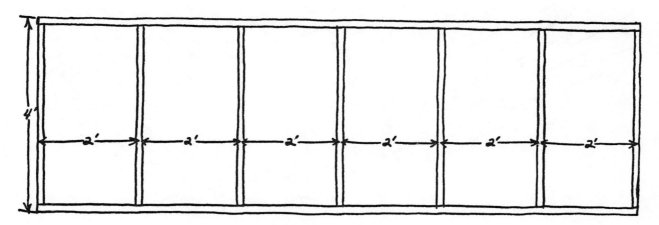

Fig. 8-3. Rear wall frame.

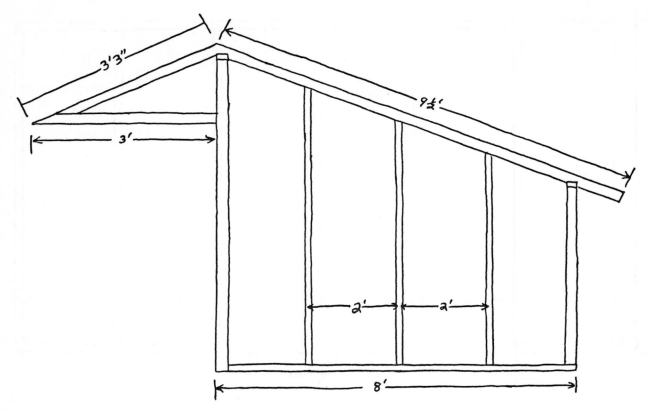

Fig. 8-4.   Side wall and rafter.

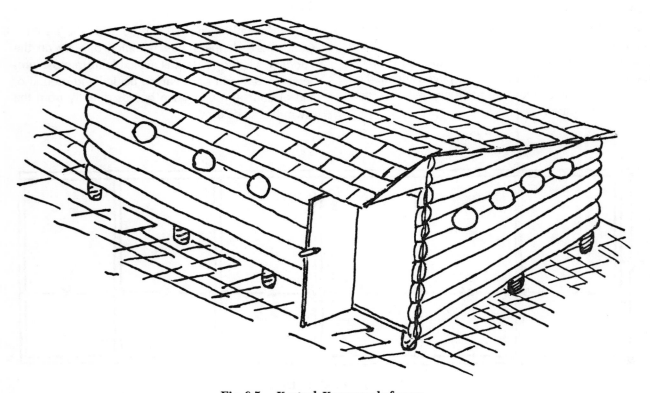

Fig. 8-5.   Kentuck Kamp ready for use.

Roof Sheathing

The roof sheathing can be ½-inch plywood. If a large snow load accumulates, it should be scraped off since the roof will not stand an extremely heavy load. The roofing can be any conventional shingle or roll roofing.

Paint

The exterior of the building can be painted a rustic brown or gray, and the inside a light green or white to make it as cheerful as possible.

Furniture

Furniture for the Kamp includes bunks and a table and benches. The bunks are made to fold up against the wall. One simple way to do this is to suspend the bunks from the walls. They are hinged in three places and a rope is knotted through a hole at each end of each bunk. The ropes are secured to the walls with large threaded screw eyes. Each bunk is a ⅝-inch plywood panel 30 inches wide and 5 feet 5 inches long. When not in use, the bunks are folded against the wall and kept in position with a screen door hook and screw eye. See Figure 8-6. The bunk should be placed within a foot of the floor to prevent little children from injuring themselves if they fall out of bed.

The table and chairs can be made as a picnic table is made. It is expected the table will be built inside the building since it will be extremely difficult to bring it in after the door and other framing is in place. The tabletop can also be used for a bunk.

Build the table by first sawing out a 3 × 5-foot panel of ⅝-inch particle board. Next, fabricate a frame of 2 × 4's with the outside dimensions the same as the hardboard panel. These 2 × 4's are placed vertically. Next, make the legs by sawing out two 2 × 4's for each end. These legs are 32 inches long. Nail them to the top and then cut out the two 2 × 6's that brace the table and project from the sides to form the seat. A 2 × 6 is also used for the seat. Study Figure 8-6. When the table is all done, it should be sanded and painted.

Beside the table a camping sink with a hose drain to the outside is convenient. Also, battery-

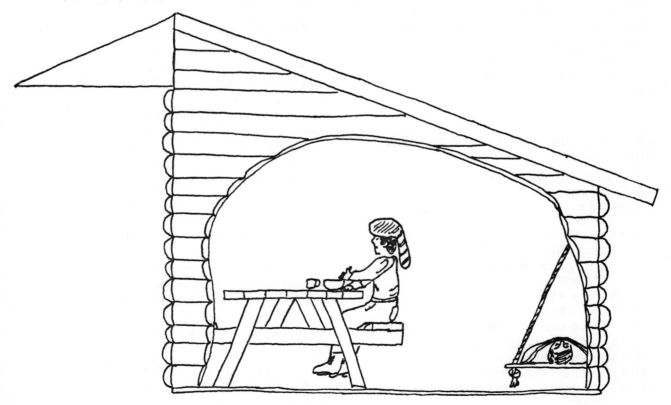

Fig. 8-6. Kentuck Kamp furnishings.

operated camping lights can be utilized to provide safe lighting so the cabin can be used day or night.

FOREST RANGER FIRE TOWER

The next project for the kids leaps ahead over a hundred years. It is called the Forest Ranger Fire Tower. In such a tower keen-eyed men have watched for dangerous forest fires for the last hundred years. They have been attacked by bears, shot at by poachers, and have often climbed down from their towers to singlehandedly put out fires that would have wiped out thousands of acres of timber if allowed to spread. Generally, the activities of these men have gone unheralded and maybe it is time we all pause for a few moments to pay them tribute.

Before building such a tower each parent should ask himself if he is willing for his child to take some risk. Also, it is to be expected that the neighborhood children will climb on it and it is possible that sooner or later one of them may fall. Injured feelings and possible litigation could be the result.

On the other hand many people have had such towers for years and no one has ever been injured on them. Further, a spokesman for the Department of Natural Resources of the state of Wisconsin asserted that no record of any injury had ever been recorded of any Forest Ranger falling from a tower or any part of it.

MATERIALS LIST FOR FOREST RANGER FIRE TOWER

1. 1 8-inch minimum diameter post, 14 feet long, for main post
2. 4 4-inch minimum diameter posts, 6 feet long, for support posts
3. 4-foot $^3/_8$-inch diameter steel cable
4. 2 6-foot 2 × 12's with $3^1/_2$ × 12-inch carriage bolts
5. 4 2 × 6's approximately 6 feet 6 inches long for platform braces
6. 5 6-foot 2 × 4's for floor joists
7. 2 6-foot 2 × 6's for closing the ends of the floor joists
8. 1 6 × 6-foot section of $^5/_8$-inch-thick particle board
9. 4 6-foot 4 × 4's for studding
10. 4 6-foot 2 × 4's for wall plates
11. 4 $33^1/_2$-foot 2 × 4's for studding
12. 3 4 × 8 sheets of $^1/_4$-inch tempered masonite or the equivalent

13. 1 12-foot and 3 8-foot 2 × 4's for the plate under the windows and the stud for the center of the window openings
14. 6 1 × 8-inch boards
15. 64 square foot of roof covering, either sawn 1-inch lumber, $^1/_4$-inch plywood, or sheet metal. Wood must be covered with roll roofing
16. 1 × 6 lumber for ladder, 6 feet of rope, nails, etc

Risks can be minimized by insisting that a definite set of rules be followed by users of the tower and violators be banished from climbing. Further, a cushion of wood shavings at least a foot thick can be provided at the base of the tower to break a fall if an accident does occur. All wagons, toys, and other objects should be kept out of the "drop zone."

Support

A fine support can be made for the Forest Ranger Tower from a single large center pole ringed by four shorter poles. The center pole should be 8 inches in diameter on the smallest end and should be at least 14 feet long. Each short post should be 6 feet long. Used posts are often available from utility companies, sometimes very cheaply. New treated posts of this type are very expensive. In many localities it is possible to purchase a live tree from a farmer or timber company and cut your own posts. Use cedar if it is available. Green posts should be peeled and allowed to dry out for at least six weeks. Then they should be treated with a good preservative. A usable container for soaking posts in preservative can be made by wrapping a sheet of plastic around the post and filling the container thus formed with the preservative.

The recommended preservative for this type of application is the toxic, water-repellent preservative. This type of preservative will make a post nearly impervious to water if it is soaked for the optimum time.

Softwood such as pine can be successfully treated in twenty-four hours of soaking. Cedar and redwood should be soaked for six days. All of the shorter posts as well as the center posts should be treated to the depth that they will be buried in the ground.

Raising the posts should be done according to a definite procedure. First the hole should be dug for the center post. After this post is set in place, each one of the shorter posts should be dug in one at a time. The center post is buried 4 feet; each short post 2 feet. Before the center post is set up, a flat surface 11½ inches wide should be sawn on its opposing sides. If it is possible to secure the services of the local telephone or electric light company to help set up this center post, fine. If not, and you have to go it alone, proceed as follows: First dig the hole and position the pole so the butt is within a foot of the hole. Next, start at the small end of the pole and raise it with a tractor bucket or with people power until it starts to slide into the hole. Then raise it a little at a time and prop it up. Exercise care so that it doesn't push too much dirt down into the hole. Once the pole slides down into the hole, plumb it up with a level and backfill around it. Then place each of the short posts, and backfill and tamp them in very well.

After all of the posts are in position, they should be joined together with a cable. Use a ⅜-inch diameter steel cable, and wrap it around the shorter posts 1 foot from the top. Use cable clamps to fit each end of the cable to the eyes of a turnbuckle, and tighten the turnbuckle to draw the cable tight. Do this on a warm day if possible.

Platform

When the cable is tightened and the posts are securely backfilled, work on the platform can be started. Use a ladder to level up the platform.

The first two boards to put in the platform are the cross girders under the floor. They are 6-foot 2 × 12's and they are placed on a flat surface sawn on the top of the post. The 2 × 12's are secured at the center with three ½-inch bolts placed through the 2 × 12's and through the post. Additional bracing for these girders is provided by 2 × 6 members nailed to the cross girders and to the tops of the short poles (see Fig. 8-7).

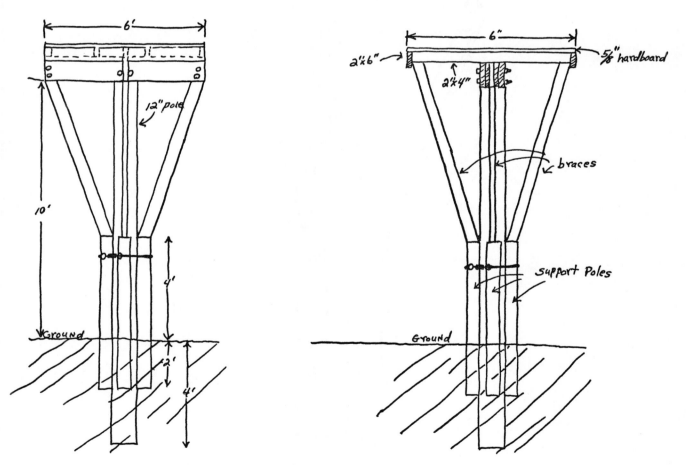

Fig. 8-7. Ranger tower platform. *Left:* Front view. *Right:* Side view.

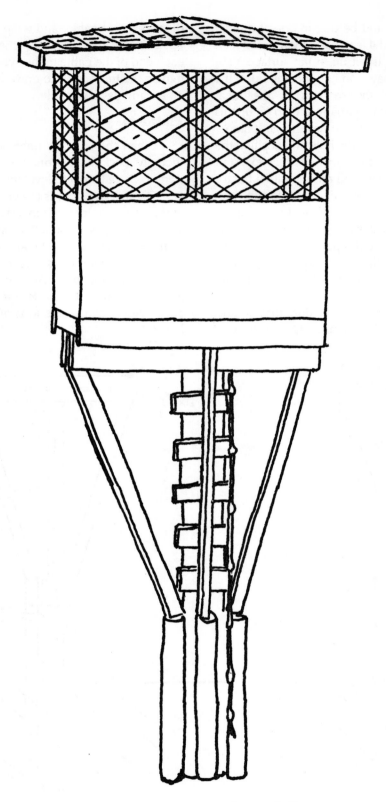

Fig. 8-8. **Ranger Tower with access ladder.**

The next structural members for the platform are the floor joists, which are 6-foot 2 × 4's nailed across the 2 × 12's. The joists are placed on 24-inch centers. A 6-foot 2 × 6 is nailed across the ends of the joists. A 2 × 6 brace is placed from the floor joists to the tops of the column poles on the side opposite to the brace for the girders. Use sixteen-penny nails for fastening the milled lumber and twenty-penny nails for fastening the 2 × 6's to the posts.

Use ⅝-inch particle board for the floor. The floor is applied before the walls are framed in so it can be used as a platform for the rest of the Ranger Tower. After the floor is applied, the entrance hole is cut in it and the ladder is installed. One way to give the children access to the tower is to use a knotted rope and shallow steps cut into one of the short posts to get to the top of the short posts. From there on up a ladder is formed by nailing boards to the center post (see Fig. 8-8). The purpose of the rope at the bottom is to discourage toddlers from making the climb.

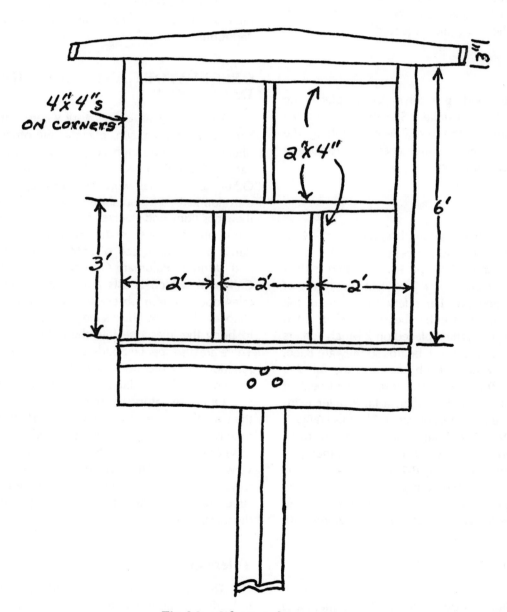

Fig. 8-9. Side view of Ranger Tower.

Walls and Windows

The walls are built so viewing is possible from all four sides. This is done by installing windows in the upper half of each wall. The lower half is solid. Start building the walls by nailing a 6 foot 4 × 4 upright at each corner. Next, toenail a 68½-inch 2 × 4 between each 4 × 4 at 36 inches height from the floor for the plate. Then the studs can be put in place. Study Figure 8-9.

The studding is 34½-inch 2 × 4's installed on 2-inch centers. It is toenailed to the floor and through the plate into the end grain of the studs.

Sheathing

When the studding is all nailed in place, the sheathing should be installed. Sheathing can be ¼-inch hardboard or the equivalent. Installing the sheathing at this time will strengthen the structure so the roof work can proceed.

Roof

After the sheathing is complete, nail a 2 × 4 with the width vertical around the perimeter of the tower outside of the top of the 4 × 4's. One stud is then placed under the perimeter 2 × 4 at the center of each window opening. For safety and comfort, the window openings can either be screened or fitted with windows. The rafters placed on 24-inch centers are formed by sawing 1 × 8 boards so they are 8 feet long, 3 inches wide at each end, and the full 7½ inches wide in the center. This forms a roof peak at the center of the tower. Further, a false rafter is installed alongside each outside rafter. The false rafter is held in place by a 1 × 4 nailed across the ends of all rafters. The roof sheathing can be either ¼-inch plywood or sheet metal (see Fig. 8-10).

This completes the Ranger Tower. It can be painted a bright color if desired or it can be made to blend into the landscape by painting it green with brown trim.

Furniture

The furniture inside the tower should include a center pedestal for the "instruments" and at least one wall should boast a topographical map, fitted with thumb tacks and strings for cross sightings on fires. Of course a snack table and chairs should be included also, because even Forest Rangers have to eat.

While many children will find the Ranger Tower an answer to their fondest dreams, perhaps your family has a child who likes to dig into things a trifle more complicated. This child may be fascinated by space, heavenly bodies, and other things celestial. If that same child happens to like photography, the next project will be very worthwhile because it is a combination astronomy observatory and photography darkroom.

OBSERVATORY AND PHOTOGRAPHY DARKROOM

The idea of an observatory where astronomers can study the stars, sun, planets, and other celestial phenomena has a rich history. In fact, the first observatory was built in A.D. 647 near Kyungju, Korea. Other observatories were built to prepare accurate astronomical tables so ships at sea could determine their positions and courses with accuracy. Some famous observatories are the French Observation de Paris, where the speed of light was first measured, and Mount Palomar Observatory in Pasadena, California, which boasts a 200-inch reflecting-type telescope. Modern observatories are usually dome-shaped. The dome houses the telescope and revolves to bring the telescope into the right viewing plane. There are also special buildings or rooms used in conjunction with the dome which are used to house instruments and special photography equipment.

Our observatory, of course, is much more simple than anything used by a real scientist. In fact, it is merely a small building with a hinged roof. The purpose of the hinged roof is to provide an unblemished view of the sky. The walls of the backyard observatory, of course, keep all extraneous light from signs, street lights, and car lights from falling on the lens of the telescope.

Telescope

The telescopes used for astronomy are either the reflecting or the refracting type. Reflecting-type telescopes can cost millions of dollars and are not to be

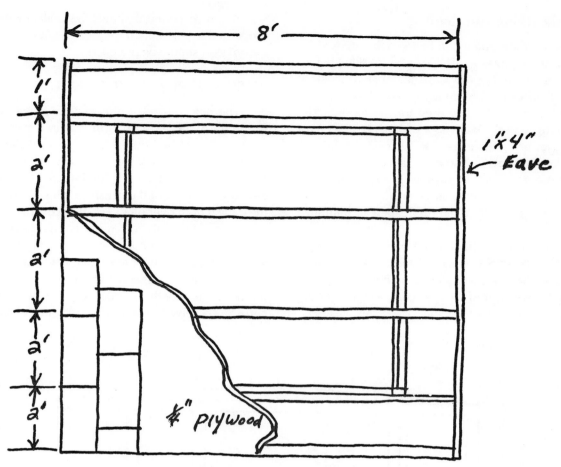

Fig. 8-10.   Rooftop view of Ranger Tower.

considered for the amateur astronomer. A suitable refracting-type telescope for backyard astronomy can be purchased for about $200.

There is adequate room with careful planning for the telescope in addition to the lights, trays, and other equipment used in the photography darkroom.

No floor is used in the observatory because a wood floor would be too unstable to use with the telescope. A thick concrete floor would, of course, be stable enough but it would raise the cost considerably.

Wall Frames

Start building the observatory by making the frame for the front wall. Cut two 2 × 4's to 7-foot, 6-inch lengths for the shoe and plate. Next, cut six 2 × 4's to 6-foot, 3-inch lengths for the studdings. Position the studs on 16-inch centers except for the two center studs, which are placed 26 inches apart to make room for the door frame (see Fig. 8-11). Nail the front frame together and put it aside until the rest of the frames are done. Each plate and sill has a tie splice. This is used for fastening the side walls and end walls together. The rear wall is made exactly the same as the front wall frame except that no door opening is provided.

The side wall frames are made with a plate and sill and studs spaced 16 inches apart. When all the wall frames are fabricated, set them up and spike them together.

A gable is nailed to the upper plate of both the front and rear wall frames. This gable is made from a 7½-foot 1 × 4 board tapered from zero width at the outer ends to the full 4-inch width at the center.

Wall Sheathing and Roofing

The wall sheathing can be ³/₈-inch exterior plywood or the equivalent. This can be installed before the roof is put on. The roof is made from two sheets of ⁵/₈-inch hardboard. The center is sealed by a rubber strip. Asphalt shingles are used for the roofing. The roof is hinged at the walls so it can be pivoted out of the way for using the telescope. One side of the roof has a cleat fastened to it so it overlaps the other. This cleat creates a seal by means of a rubber strip fastened to it. Reinforcing strips of 1 × 4 boards are fastened to the inside of the roof panels to minimize warping.

No windows are built into this observatory since they would be unnecessary. Water and electricity is supplied to the observatory from the house. A garden hose and a waterproof electric cord will supply the utilities with little expense.

A simple pedestal should be built for the telescope and a sink, bench for the trays, and a table for the enlarger should be provided.

MISSISSIPPI RIVER BARGE

The next project in this chapter comes straight from one of the most colorful eras in American history: the day of the Mississippi riverman. This is not to say that there is no cargo being hauled on the "Big Muddy" today; it is still an important waterway. However, the day of the swashbuckling rivermen and the danger from river pirates is all but over.

One of the most colorful of the rivermen in those days was the river pilot. The services of good pilots were so eagerly sought after that they were afforded every measure of respect, and they often received as much for a month of their services as a skilled carpenter received for a year's work. Such an occupa-

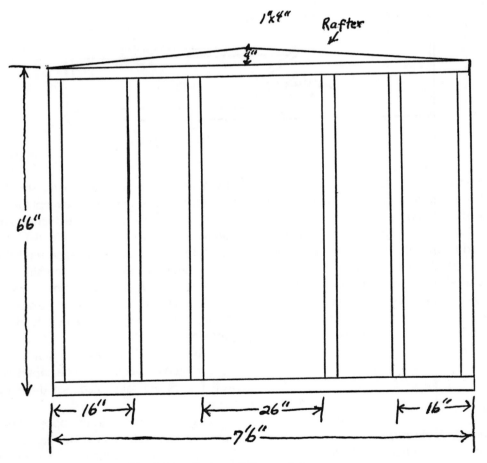

Fig. 8-11.   Front wall frame of observatory.

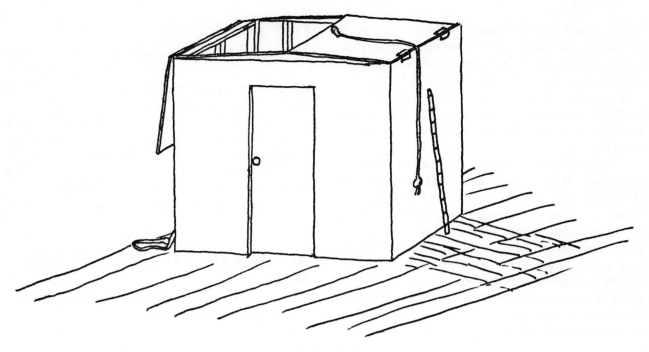

Fig. 8-12.   Backyard observatory.

tion was bound to attract some unusual people and it did. One of the most outstanding was a young lad named Samuel Clemens who in later life received worldwide fame as a writer under the pen name of Mark Twain.

This brings up a story that has been handed down in my family since the days of my great-grandfather, who was a Mississippi riverman, decorated Civil War veteran, and rugged independent backwoodsman all of his life.

Grandfather Jim was a deckhand and general roustabout on the cargo barges. He said a deckhand was as far removed from a pilot as a private from a general, and, in fact, he very seldom saw the pilot. Therefore, when Grandfather Jim saw a young fellow lounging about on the aft section of the barge he sought to draw him into conversation. After an exchange of pleasantries Jim asked the fellow where he was from. When the stranger replied "Boston," Jim immediately assumed he was a paying passenger and a "Yankee traveler," green to the river and probably heading further west. Always ready for a joke, Jim took the fellow for a tour of the boat, deliberately attaching the most ridiculous names to parts of the

barge. After he had completed the tour he kept the stranger buttonholed for the better part of a hour spinning completely untruthful tales of life aboard the ship: attacks by ruthless pirates, shipwrecks, and the horrible food and working conditions that the deckhands were forced to put up with. For good measure he fabricated a few stories about the beautiful women waiting in the little towns along the way. The stranger listened very intently to every word, showed him the greatest respect, and thanked him very intently and gratefully when Jim exhausted his supply of stories and took his leave.

Grandfather was so tickled by the fact that the stranger swallowed every word he said that even before the stranger walked away he felt such a knot of laughter rising in his chest that fighting it back actually strained a muscle in his side. He could hardly get out of his bunk the next day and, in fact, it was two days before he could get around well. When he did, he went looking for the stranger after his watch one afternoon.

As he walked by the wheelhouse he happened to glance at the pilot, mostly to see whom they had picked, since they had changed pilots at St. Louis two days before. He could scarcely believe his eyes when he saw the "Yankee traveler" standing all alone at the wheel, guiding the huge craft through the snag and whirlpools of the river. The outrageous lies he had told actually fell on the ears of the pilot himself. Probably no one except the immortal Mark Twain himself would have let the deckhand play out his hand that way.

When Grandfather Jim finally realized what had happened, he slunk away and never ventured near the wheelhouse again. When he got to Prairie du Chien, he collected his wages, went back to the farm for an extended period, and never, never worked a barge that had a pilot named Clemens again. Years later when Twain wrote his famous book *Life on the Mississippi* he mentioned the incident but, remarkably enough, Clemens apparently never mentioned it to a single soul otherwise and no one on the river ever knew exactly whom he was talking about.

Laying Out the Barge

Start building the River Barge by laying it out with stakes and strings. This gives you a chance to visualize it in a given location. The Mississippi River Barge is 8 feet wide and 16 feet long. It should be set on posts placed at 2-foot intervals across the width and at 4-foot intervals along the length. It is expected each post will project 18 inches above the ground at the lowest point. The posts should be at least 3 inches in diameter and they should be treated to prevent decay. Fence post sections will do nicely. See Figure 8-13.

MATERIALS LIST FOR MISSISSIPPI RIVER BARGE

1. 13 4-inch diameter minimum poles for foundation
2. 6 8-foot 4 × 4's for girders
3. 7 8-foot 2 × 4's for floor joists under pilot house

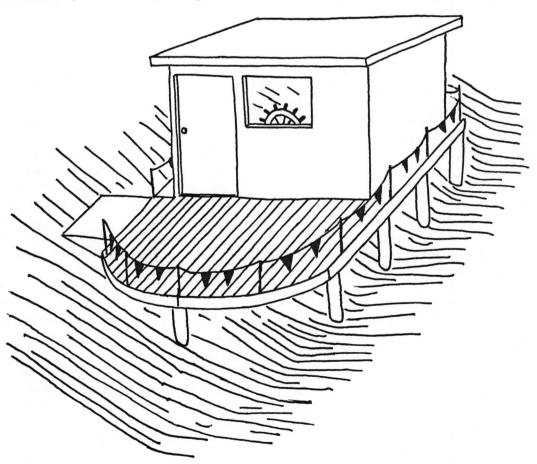

Fig. 8-13. Mississippi River Barge.

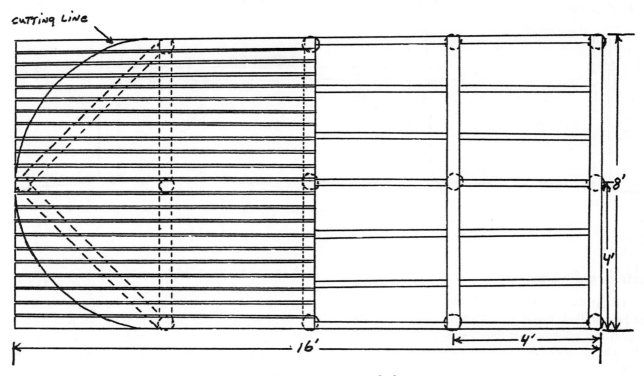

cutting line

Fig. 8-14. River Barge deck.

4. 21 8-foot 2 × 4's for decking
5. 2 4 × 8-foot sheets of 5/8-inch particle board
6. 10 8-foot 2 × 4's for pilot house studding
7. 4 10-foot 2 × 4's for roof rafters
8. 7 4 × 8-foot sheets 3/8-inch exterior plywood
9. 65 square feet roofing
10. Pilot house window, wheel and inside furnishings. 1-inch dowel, rope, etc.

Start laying out the barge at the lefthand rear corner. Drive a tall stake there and project all the rest of the measurements from it. The first line to project will be the lefthand side of the barge. This line will be 12 feet long. Find it by tying a chalk line to the starting stake, pulling it tight, and tying it to another stake about 13 feet from the starting point. At 4-foot intervals along this line drive stakes to mark the locations of the posts in the lefthand side of the barge. The four outline posts will form a line 12 feet long. The center row of posts will be 16 feet long. See Figure 8-14. When the lefthand row is outlined, go back to the lefthand rear corner and use a carpenter's framing square, the chalk line, and a measuring tape to find a

point at right angles to the lefthand rear post which will mark the 8-foot width of the barge. Drive a stake there. Then go to the front stake of the lefthand line and project the front stake of the righthand line using the same methods. All that remains then is to measure off for the posthole locations.

The posts should be set into firm soil and tamped very well. This barge is well adapted to being built on a slope or hillside. In fact, that is where it should be built. Having a steep dropoff on one side enhances the excitement of using the barge since the "riverman" can pretend this is deep water.

If a slope is used, of course, some posts will be much longer than the minimum 18 inches. Sawing off round posts so the tops are all level is handily done by making a temporary jig for guiding the sawblade. The jig is simply two pieces of 1-inch board nailed at right angles to the post. The boards are nailed right on the cut-off mark. Care is taken to nail them on level so they will guide the saw in making a level cut. See illustration 8-15. Use a coarse-toothed hand saw or a wood saw.

End View   Side View

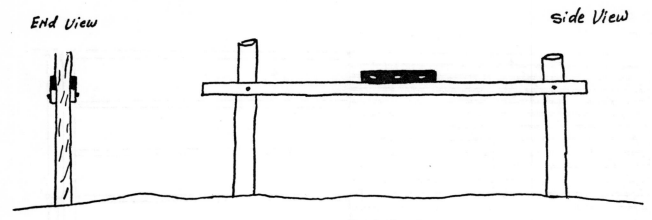

Fig. 8-15.   Squaring the tops of posts.

Installing the Decking

When the post tops are all leveled, the decking can be installed. The rear end (stern) of the barge will support the pilot house; thus the stern of the barge will include floor joists covered with flooring.

However, the first construction stage is the same for the entire length. Spike 8-foot 4 × 4's on the tops of the posts across the width of the barge. These will be the girders. The bow section has a 4 × 4 extending from each side to the single post in the center row (see Fig. 8-14).

When the girders are all in place, nail 2 × 4 joists on 16-inch centers on the 6 × 8-foot section that will be the floor of the pilot house. Cover the joists with ⅝-inch particle board or the equivalent. All of the rest of the deck will be covered with 2 × 4 boards. They will be separated by ½-inch cracks to allow rain to fall through. The decking in the bow section is allowed to extend past the girders and is finally cut on a radius to form the bow (see Fig. 8-14). This radius can be marked by stretching a string from the center of the barge, 4 feet from the extreme end of the bow.

Pilot House

After the deck is complete, the pilot house can be made up and put in place. It measures 6 × 8 feet and is equipped with a "steering wheel," two bunks, and even a tiny galley. This allows the barge to be useful even on rainy days. The door opens onto the deck so quick action can be taken to dispel "pirates" or other undesirable boarders. A few barrels or boxes can be placed on the deck to simulate cargo.

Start building the pilot house by first fabricating the framing for the front wall. The shoe is 6 feet long, and the plate 7 feet in length (see Fig. 8-16). The studdings are spaced on 24-inch centers. The door frame is 24 inches wide. The overhang of the plate is used to tie the front and side walls together. The door, located on the righthand side of the front wall, is fabricated from a ¾-inch-thick plywood panel 23 inches wide and 72 inches high. A window is located in front of the wheel (see Fig. 8-13).

The rear wall plate is 7 feet long and the shoe is 6 feet long (see Fig. 8-16). The studding is also placed on 24-inch centers. The roof rafters are 9½-foot-long, 2 × 4 boards spaced on 16-inch centers. The side studdings are placed under the roof rafters (see Fig. 8-17), marked accordingly, sawed, and nailed in place. The sheathing for the side walls and the roof is ⅜-inch exterior plywood or ¼-inch tempered masonite for the sides and ⅜-inch plywood for the roof. The roofing can be asphalt shingles.

When the outside is done, the railing around the deck can be made up and installed. The railing is simply a ⅜-inch diameter nylon rope strung through 1-inch diameter wooden dowels set into the periphery of the deck (see Fig. 8-13). Each dowel is drilled for the rope before it is set in place. Brightly covered streamers hung from the rope impart a festive appearance.

Inside the pilot house a wheel and wheel stand is

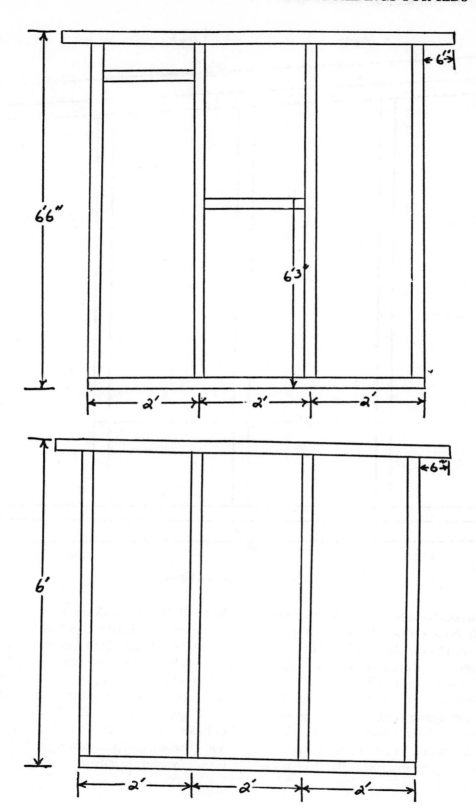

Fig. 8-16.   *Top:* Front frame of pilot house. *Bottom:* Rear frame of pilot house.

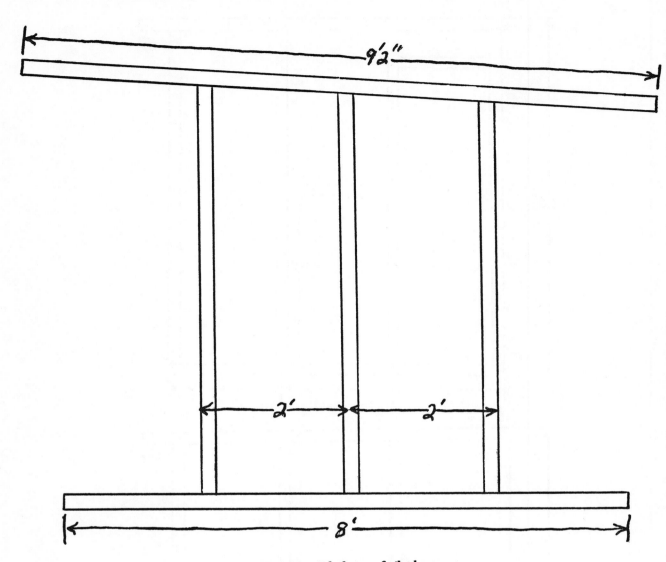

Fig. 8-17. Side frame of pilot house.

built. Build the wheel by setting 1-inch wood dowels into a hub made from a 3½-inch section of 4 × 4. Pivot the wheel on a ¼-inch bolt anchored to a 4 × 4 stand. An automobile steering wheel also can be utilized if you do not desire to build your own.

Use your imagination to create your galley. One way is to use a camping stove for cooking and a plastic camping sink which contains its own water for washing dishes and hands. By all means use the traditional sailor sleeping arrangement, the hammock.

Suspend the hammock from eye bolts into the studding of the width of the pilot house.

Although the Mississippi River Barge is a lot of work to build, it will be very useful. Besides a place for the children to play in good weather or poor, it will provide extra sleeping quarters when there are house guests, turn an ugly unused slope into a place of beauty, provide an outdoor eating deck for adults and children, and generally make the homesite more interesting and attractive.

Chapter 9

SAUNAS, SCREENHOUSES, AND OUTDOOR FURNITURE

SAUNAS

When we moved from southern Wisconsin to extreme northern Wisconsin we encountered a very different lifestyle from the one we were accustomed to. We found many things interesting and delightful, but one of the most intriguing was the sauna baths and the way the natives in the Upper Peninsula of Michigan and in the northeastern corner of Wisconsin made and used their saunas.

The sauna bath is a Finnish custom, and it no doubt came to America with the thousands of Finlanders who immigrated to this country around the turn of the century. Local historians say that the immigrant Finlanders built their sauna house first and lived in that until they could get a permanent house built. This gives some indication of the reverence that the Finlander holds for his sauna bath.

A sauna bath is largely a method to induce wholesale perspiration. Unlike a Turkish, or steam bath, the conventional Finnish sauna subjects the body to very high temperatures and a very low humidity. This causes the pores of the body to open and literally gush forth perspiration. As any doctor knows, induced sweating is the best method known for cleaning the skin.

A sauna bath does much more then clean the skin, however. It eliminates colds and relieves rheumatism, muscle spasms, and pain from overtaxed muscles. It "boils" away blackheads and pimples and imparts a healthy glow to the skin. It has an important psychological effect also as it relieves nervous tensions, pressure, and preoccupation. It is said no one can come out of a sauna bath in ill humor. Personally, it makes me feel like a new man and I can easily recharge myself with a sauna bath so that I feel as good after a day's work as before I started, not to mention that it is the fastest, easiest way I know to lose from one to three pounds.

However, no one should go into a sauna right after an extremely heavy meal. Generally speaking, a person troubled with heart disease, high blood pressure, or respiratory ailments should avoid the high heat of the sauna bath. It also should be avoided immediately after heavy exertion. Many people report very negative effects from the sauna after consuming alcoholic beverages. However, they also report that a sauna has a very beneficial effect on the "morning after" blahs.

Since the sauna is a Finnish custom and the Finns have been refining it for about 2000 years, their customary way of enjoying it should be followed. First, build a fire in the sauna stove and heat the room to at least 140°. While the room is heating, find or cut a bundle of birch twigs about 2 feet long and tie them together. This bundle of twigs is called a "whisk," and it is used for beating or whisking the body. More about this later. Then arrange for a container of hot water, cold water, and soap to be available in the sauna house.

Enter the sauna naked or with a loose towel draped around the waist. Sit on the lowest bench at first. When you become accustomed to that, move to the highest level bench in the sauna and when that becomes enjoyable, increase the heat by building up the fire until a temperature of at least 165° is reached. When the heat is as high as is comfortable, lie down in the prone position on the upper bench and elevate the feet. Maintain this position as long as is comfortable, which is about 15 minutes for many people. This first phase is called the perspiration phase. Follow this by a roll in the snow or dip in the lake if possible to cool down. A cold shower will also work, as will pouring cold water over your head. After the cooling off period rest a short time and then return to heat. Repeat this as often as you feel the need. Some hardened sauna users do this five times or more.

After the perspiration stage comes the steam phase. Steam is produced by simply sprinkling water on the stone, either by dipping the whisk in the cold water container and then sprinkling that on the stones or by using a long-handled dipper to pour some water on the stones. This should be done while lying down if possible and for that reason the sauna should be arranged so the water can be reached from the top bench. A minimum of water should be used so the steam does not become too stifling. After steaming for about five minutes, enter the whisking stage.

Whisking is beating the body lightly with a bundle of birch or cedar twigs. This loosens the old skin softened by perspiration and stimulates blood flow to the capillaries. Whisking starts at the upper body and progresses to the extremities. If you do not desire to use twigs, somewhat the same effect can be achieved by a brisk towel rubdown or scrubbing the body with a long-handled shower brush. Some people think a twig whisk is too much trouble as twigs do tend to dirty the floor of the sauna. Also, traditionally the birch twigs are cut in early summer when the leaves are on. They are then dried and kept for use as needed.

After the whisking comes the bathing. First, put hot water and a little soap in a container such as a wash basin. Stir this around until it makes a foamy lather. Then take handfuls of the lather and cover the body with it like a shaver lathering his face. Next, take the whisk and scrub the body with it very well. Then completely rinse off the soap with clear warm water. Take care to rinse off all the soap and also rinse the benches and stools and floor. After the washing period return to the perspiration stage for a short time to get the body very hot again. Then whisk for a short period again.

The next stage is to cool off very quickly. Finlanders do this by rolling in the snow or by jumping into a cold water lake or even by jumping into a hole in the ice. American sauna users take a cold shower.

The drying stage follows the cooling off period and if it can be done outside in the air, that is fine. Generally, the hair and face should be dried with a towel and the rest of the body air-dried. If it is necessary to dry inside, use a warm, dry room, not a warm, moist one since you might never get dry in a moist room. In the city some people use alternate hot and cold showers for cooling off, then dry with a towel.

Just as the sauna bather lies down when he is perspiring, he should lie down when he finishes the sauna. A rest of ten to fifteen minutes neither speaking nor thinking is recommended. After that the bather can dress and the sauna is complete. He will then be very thirsty and if he desires to gain back the weight he has lost he can drink copiously for a few minutes. Many of the native saunas were followed by a meal. This meal sometimes could be cooked right on the sauna stones while the bath was taking place. Avoid using odoriferous foods or foods which have to be boiled if you cook in the sauna, since they can affect the humidity and also spoil the clean air of the sauna.

All this is a long introduction to building a sauna, but without the tradition it just becomes another way of taking a bath and a fine legend will be cheapened.

A sauna should be built separately from the main house. This is because it is somewhat of a fire hazard, especially if wood heat is used, as many people think it should be. It also should be comfortably large but not too large to heat. Our sauna is 8 × 10 feet.

Drainage and Foundation

Start building the sauna by building the foundation. Care should be taken that the drainage is very good from the foundation of the sauna since water will run down into it each time the bath is used. If you are building in a suburban location, the drain will have to be connected to the septic system or the sewer system unless you have a water recycling unit. Very likely the local plumbing inspector will have some regulations concerning this if your property is zoned. In many states now the buildings must be built back a specific distance from the lake, river, or stream. Consult the local zoning administrator. In areas where field tile is used for drainage, the water from a sauna probably could be run directly into the tiles if soap remover is used in conjunction with the bath. A simpler way is to excavate the entire length and width of the foundation and fill the excavation with coarse gravel. This creates a leach bed which will easily take care of the small amounts of water used with a sauna.

MATERIALS LIST FOR SAUNA

Sauna foundation and floor
1. 15 6-inch diameter poles, 3 feet long
2. 8 1-inch diameter 12-inch length plastic pipes (optional)
3. 1½ yards of 2-inch gravel (optional)
4. 5 4 × 4's 8 feet long
5. 6 10-foot 2 × 6's
6. 5 4 × 8 sheets ⅝-inch particle board

Framing
7. 16 8-foot wall studs
8. 4 10-foot wall studs
9. 2 8-foot 2 × 4's for shoes

10. 2 10-foot 2 × 4's for shoes
11. 4 10-foot 2 × 4's for wall plates
12. 1 6-foot 2 × 4 for door and window frames
13. 14 8-foot 2 × 6's for rafters
14. 4 10-foot 2 × 6's for ridgeboard and cross beams
15. Sheathing to cover approximately 400 square feet
16. Siding to cover approximately 240 square feet
17. 160 square feet of roofing
18. Window and door
19. 18-foot 1 × 4 lumber for screen door

The actual bed for the building can be cedar posts placed into the ground or a foundation of cement blocks. Posts are preferable in some areas because of uneven ground or hard digging conditions. When posts are used, they should be placed on 30-inch centers on the perimeter. Three rows of posts should be used to provide 4-foot spacing for the 4 × 4 floor girders. See illustration 9-1. The floor joists can be placed on 16 inch centers, and they should be 2 × 6 boards or the structural equivalent.

Flooring

Ideally, two layers of flooring will be used with a layer of bituminous felt sandwiched between them. Each layer should be ⅝-inch particle board or the equivalent. Tongue-and-groove 2 × 6's also can be used, but in this case insulation such as sheets of foam or fiberglass should be fastened underneath the floor. Drain holes should be drilled through the floor to allow the water to run into the leach bed below. If desired, each hole can be fitted with a section of plastic pipe that extends to the gravel underneath the floor (see Fig. 9-2). In this way cold air can be prevented from seeping into the building around the bathers' feet. It is usually prudent to place racks on the floor so the bathers' feet are kept off its cold surfaces. Naturally, a concrete floor can be used. In this case one drain hole connected to a drainage system would be utilized. The entire floor would then slant to the hole.

Nailing the Studding

The authentic Finnish sauna should be built of logs. However, you can achieve a log effect by the use of split log siding. The next step after building the floor is to nail the studding in place. Since it is desir-

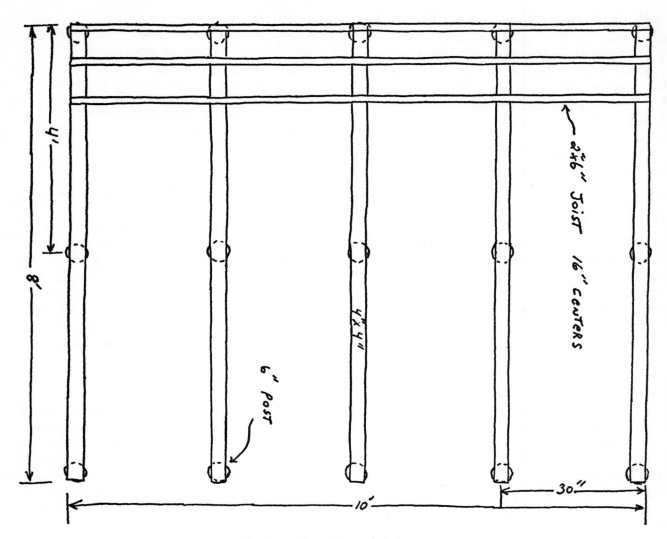

Fig. 9-1.   Floor joists and girders.

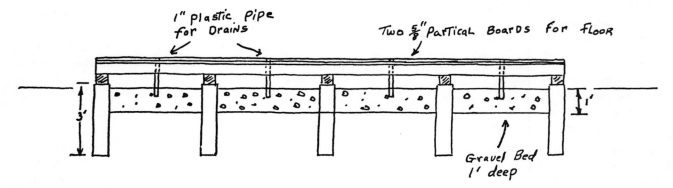

Fig. 9-2.   Side view of gravel bed and flooring.

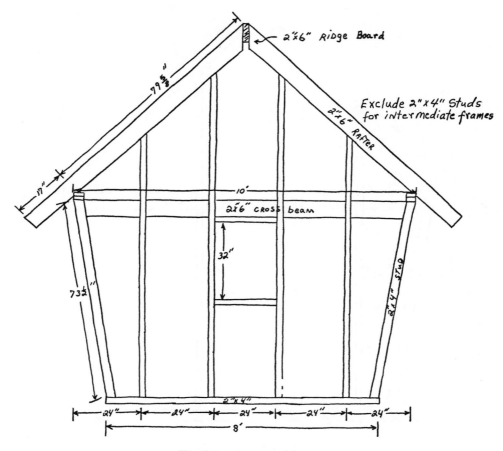

Fig. 9-3.   Sauna end frame.

able to have a flared wall, the studding will not be completely vertical. Instead, it will sit on the sills at an angle so that the top is actually 24 inches wider than the bottom (see Fig. 9-3). This angle not only imparts a dramatic unconventional look to the building, but also provides room for the top benches so they can be made wide enough for prone position bathing. All of the studs are made the same length and each has the same angle at the plate and sill. A double plate is used.

Start with an 8-foot 2 × 6. Square one end. Place the framing square at one corner of the stud so it will form a line that when it is sawed off will remove a right triangle piece with a ³/₄-inch base. This angle will of course be in opposite directions but exactly the same on each end of the stud. Mark this angle and cut it out. Check the stud to see if it is properly cut and then use it for a pattern for the rest of the side rafters. One slight deviation from normal construction

is that the angled studs have to extend clear to the outside of the back and front walls.

Door and Window

The door is framed into the front end and the window in the other (see Figs. 9-3 and 9-4). The door frame should be made of 2 × 6 material if it is available. This allows an inside and outside door to be utilized so heat can be conserved in the sauna. Any kind of commercial outside doors can be utilized if you desire to purchase them. Also two "garage entry" type doors with windows can be used to eliminate the window in the back of the sauna. Of course, if the sauna is being built in a populated area, the window glass should be stained or treated or else one door should be fitted with a shade to discourage window peekers.

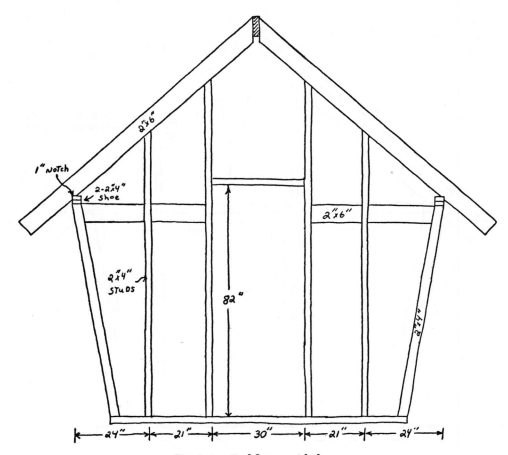

Fig. 9-4. End frame with door.

Crossbeams

Crossbeams are placed on 4-foot centers. Ideally, the will be made of 2 × 8 stock, but 2 × 6 or the equivalent in poles is also adequate. After the building is done the crossbeams can be used for bed pieces for the bathing racks. This sauna has the capacity for six bathers all at the same time while it also can be utilized by only one bather without serious waste of fuel. For wall sheathing use ship lap or ½-inch plywood and half-long siding.

Rafters

The rafters should be made from 2 × 6 stock and placed on 16-inch centers, especially in territory where the snow load is substantial. Make the pattern rafter by laying an 8-foot 2 × 6 out on sawhorses. Use 5 inches for a measuring line. Scribe this line from one end of the rafter to the other. Lay the 12-inch mark on the blade of the square and the 5-inch mark on the tongue of the square at the measuring line. Move the square five times, since the run of the rafter is half the span of ten feet, to find the length of the rafter. Another way is to multiply five times the distance between the 12-inch mark of the outside edge of the blade of the square and the 10-inch mark on the tongue of the square. This distance is 15⅞ inches. Multiply this by the span of 5 feet and we have 5 × 15⅞, or 79⅜ inches, which is the total length of the rafter. After finding the correct ridgeboard angle of the rafter, a measurement from both points of this angle will reproduce the wall plate angle at the desired position on the rafter. This, of course, simplifies making the rafter and minimizes the chance for error.

In this rafter the heel or tail is left full width and the rafter is notched for the wall plate. A ridgeboard is used. Thus, after the rafter is made, saw half the ridgeboard width from the length of the rafter at the ridgeboard cut. This is ³/₄ inch. Now as long as you make each succeeding rafter just exactly the same as the first one by laying the first one on the succeeding rafter boards, no problems will appear even if some slight error has been committed in laying out the pattern rafter.

The ridgeboard will have to be placed on a temporary stand until a few rafters are nailed to it. This stand should project above the wall plate approximately 53³/₈ inches, although slight adjustments may be necessary to fit the angle of the rafters exactly.

After the rafters are in place, set the gable studs in place on 16-inch centers and mark and cut them. Then apply the roof sheathing.

Roofing

The roof material can be roll roofing or composition or shake shingles (see Fig. 9-5), but at least two layers of felt should be applied under the roofing. This minimizes the need for insulation under the roof. However, if insulation is desired, use 6-inch insulation in the ceiling and 4-inch in the sides. Then finish the inside with solid boards or paneling which will withstand heat and humidity. One-quarter-inch outdoor plywood treated with a water preservative should also be satisfactory.

Cold and Hot Water Barrels

After the shell is done, the inside furnishings can be built by hand or they can be purchased from commercial sources.

In a sauna that can't be equipped with running water some water storage facilities must be provided. Generally this consists of two large 20 to 30 gallon containers. The economical containers of this size are generally used barrels. It works very well to cut the tops out of the barrels and equip them with a wooden cover. The barrel which will be the hot water barrel is positioned next to the stove and equipped with pipes that project from the bottom and the top of the barrel through the stove (see Fig. 9-6). Several refinements can be incorporated into this water heating arrangement but good results can be obtained by simply running a copper pipe through the stove. The ends are connected too through the stove and slightly separated by one being higher than the other in the barrels. This creates a circulation caused by the heating of the water in the pipe. In fact, this water turns to

Fig. 9-5. Finished sauna.

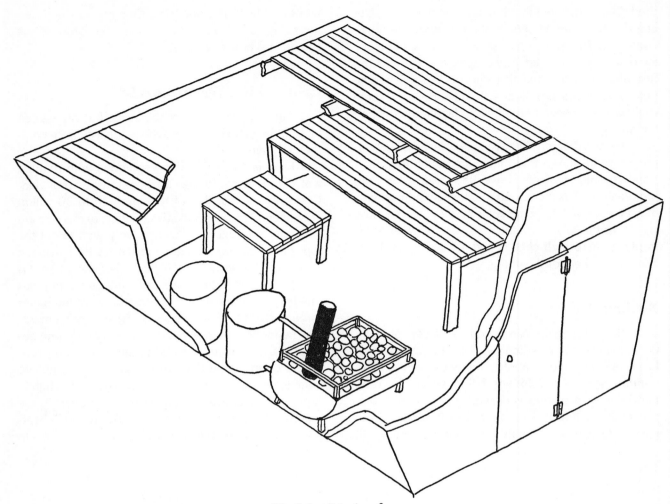

Fig. 9-6. Interior of sauna.

steam and bubbles out the top. Cold water then comes in to take its place and is in turn heated. The size of the pipe in the stove largely determines how fast the water will heat.

Barrel Stove

Since we designed and used a barrel stove especially for our sauna, a brief description of how we did it may help a prospective builder.

First, take a 55-gallon steel drum. Cut it in half lengthwise with a cold chisel. Use the section with the large hole. Then go to the blacksmith and obtain a sheet of ¼-inch boiler plate measuring 25 × 36 inches from one end. Have the blacksmith cut a hole and weld a 6-inch pipe onto the boiler plate to use for a stovepipe collar. Further obtain 10 feet of 1-inch

angle iron and bolt it to the underside of the boiler plate so it will fit inside the barrel half. This is to secure the plate to the barrel. This forms the top.

Now have the blacksmith braze a large U-shaped (4-inch) water pipe to the barrel shell so that it lies just under the top of the stove. Make sure the ends are reduced so that it can be fitted with ¾-inch well pipe.

Next, make the door for the stove. Take another 8½ × 11½ piece of boiler plate and cut out an opening in it 6½ × 9½. Next, take a piece of ⅜-inch plate steel and weld a piece of ½-inch metal tubing to it to use as a hinge. Then weld two pieces of matching tubing to the boiler plate frame so that a metal rod can be placed through them to complete the hinge. The frames of boiler plate are bolted to the barrel and the plate steel door is hung from the hinge. The latch is a piece of 1-inch strap iron that pivots with a bolt. A small piece of angle iron is then welded to the barrel and notched so the strap iron latch will fit in it. Cut the opening for the door at the end of the barrel with the large bung. The stand for the barrel stove is also made from angle iron. Further, 3/4-inch pipe fittings must be brazed to the water barrel so the circulation pipes from the stove can be fitted to it. Connecting pipes are then made up which will join the water barrel and the stove. Copper tubing, although expensive, works well for this application but galvanized water pipes can also be used. A large dipper is used to remove the water from the barrel or the water barrels can be fitted with faucets by simply brazing the fittings to the barrels. The cold water barrel is placed beside the hot water barrel. After the stove is made up, a large container such as a section of barrel can be placed on the top of the sauna stove and filled with rocks. When they are well heated, water is sprinkled on them to create steam for the steam phase of the sauna bath.

Besides the stove and water pots, the racks are generally handmade. They are generally made from aspenwood if it is available since it is a poor conductor.

If it is not desirable to make up an automatic water heating unit, just sit a water container on the top of the stove. Generally only a gallon or so of hot water will be needed for the bath. Too much water and steam in the sauna could raise the temperature too high.

Three separate platforms at different altitudes can be used. The bottom will be used for bathing, the center for the whisking platform. Study Figure 9-6 for the proper method. Also a partition should be included to form a dressing room.

How to Convert a Room to a Sauna

Of course, not everyone lives in the country where they can build a separate sauna. This is fine also. Most large homes have a room that can be converted to reproduce the sauna conditions. A closet for instance will work fine. Just insulate all the walls with at least 3 inches of insulation. Install a bench and an electric heater with a controllable thermostat. Enter the room, build the heat as high as you can stand it and perspire away your problems. It

helps, of course, if you can reach the shower from the closet without parading through the living room or some other room where there is apt to be people. It is a good idea to plan your new house with a sauna room included. In that case it can be adjoining the bathroom. A single person or even a couple could use their bathroom for a sauna, but this of course wouldn't be too convenient if there was one bathroom and several people.

People who have a swimming pool and a pool room have almost the ideal artificial setup for sauna bathing. Just build a small insulated room off the pool and "sauna" away. Commercial units are also available for this type of installation if you don't desire to build your own. Be sure to have a room large enough to lie down in while sauna bathing, though, since this is part and parcel of the sauna.

SCREENHOUSE

One of the most delightful buildings to have in proximity to a sauna is a screenhouse that a bather can relax in after the rigors of the sauna. It is still better if the screenhouse can be made safe from prying eyes so the bather does not have to don clothes before entering the screenhouse. A screenhouse is, of course, a fine place to relax in after work or play or a place to eat outside, sleep outside, or do a dozen other outdoor things. This screenhouse is unconventional in design and very useful to a dozen different lifestyles.

The screenhouse described here is 6 × 9 feet, which is large enough for a double bunk, a card table and chairs, or a custom-made trestle table and benches. It is most suitable for a family of about four persons. If more room is desired, of course, the screenhouse can be made larger. The roof is made overlarge so it will catch rainwater, which is the best for sauna bathing. A catching basin and pipes could be used to direct the water from this roof to the cold water container inside the sauna. A simple filter could be provided on the outlet from the catching basin to strain the dust and other impurities from the water. The large roof also serves as a shaded area when the screenhouse will be used for other than sauna use. It is expected that the inside of the screenhouse will have a wood floor. Generally it is advisable to set the screenhouse permanently in one place.

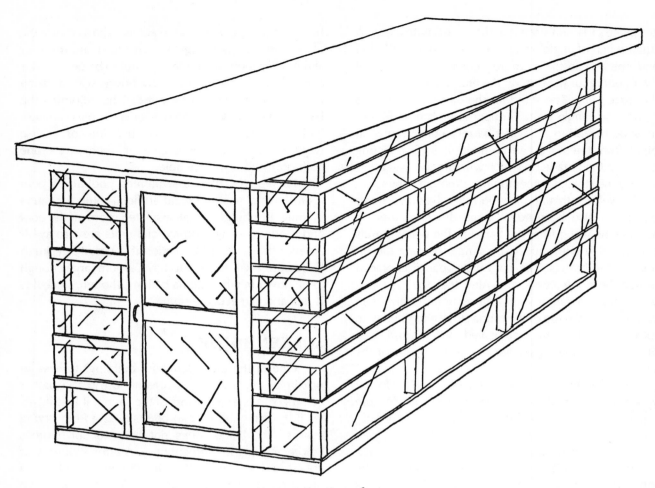

Fig. 9-7. Screenhouse.

MATERIALS LIST FOR SCREENHOUSE

1. 10 8-foot 2 × 4's for side and end frames
2. 8 6-foot 2 × 4's
3. 1 9-foot 2 × 4 for floor. 3 12-foot 2 × 4's for rafters.
4. 8 6-foot 2 × 4's for floor and roof
5. 54 square feet ⅝-inch plywood
6. 54 square feet ⅜-inch plywood
7. 60 square feet roofing. Nails, etc.
8. Approximately 200 square feet screen
9. 4 pounds ten-penny nails, 2 pounds ½-inch roofing nails

Floor

Start by building the floor. Cut two 2 × 4's 108 inches long and four 2 × 4's 72 inches long. Use ten-penny nails to nail the 72-inch 2 × 4's between the 108-inch framing members to form the floor frame (see Fig. 9-10).

When the floor is framed in, nail tongue-and-groove 2 × 4's or ⅝-inch plywood to the framing. This flooring should extend to the outside of the frame. After that, the studding is toenailed on the top of the floor. The studding is positioned on 36-inch centers (see Fig. 9-11). The front studs are 7 feet long; the rear studs are 78 inches. Note that the door frame is 3 feet wide (see Fig. 9-12).

Roof

The roof framing is 2 × 4's nailed across the studs on the outside. They are 12 feet long with an overhang of 18 inches both front and back (see Fig. 9-11). A center 2 × 4 rafter also is used (see Fig. 9-13).

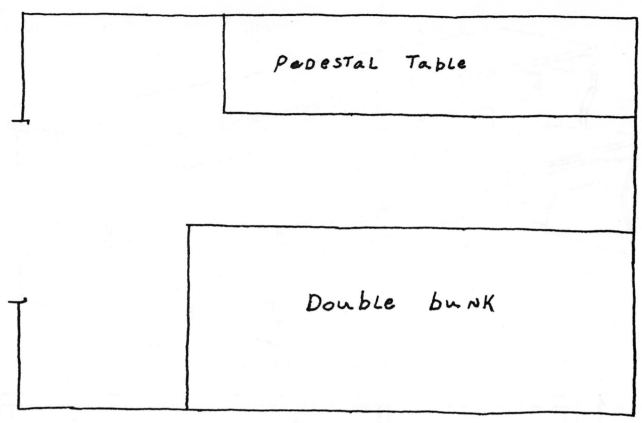

Fig. 9-8.   Floor plan for screenhouse.

Additionally a 2 × 4 plate is used across the tops of the width studs (see Fig. 9-12). The roof should be covered with ³/₈-inch plywood or the equivalent. A smooth roofing such as noncorrosive metal should be used for the roof if it is desired to keep the water clean. If this screenhouse is not to be used for this purpose, a more economical roofing such as roll or composition roofing can be used.

Screen

The space between the studs is slatted with 1 × 3 furring strips of lumber. The furring strips are placed 1 foot apart. They are toenailed to the studding with six-penny finishing nails. When the slats are nailed in place, they provide shade as well as a base for nailing on the screen. Either plastic or metal screen should be used for this application. If this screenhouse will be used without the sauna, regular metal screen will be most satisfactory.

Screen Door

A standard commercially built screen door can be used. However, it is not difficult to build your own with a 1 × 4 lumber frame. Since nailing this type of corner is troublesome, a joint called a mortise and tenon joint must be utilized. This involves sawing a slot in the end of a board and creating a tenon joint on the end of another board to fit into the slot (see Fig. 9-14). A simple cross joint must also be used for the center brace.

Both joints can be formed with a hand drill and a hand saw. To make the mortise section of the joint, drill a hole ¼ inch in diameter through the side framing members. This slot should be the same distance from the end of the piece as the side panels are wide, generally 3½ inches. Next make saw cuts to form a slot ¼ inch wide. Also saw the connecting board to form a male tenon ¼ inch wide to fit in this slot. This joint will be used at all four corners. The crosspiece is joined to the sides by cutting a square notch ³/₈ inch

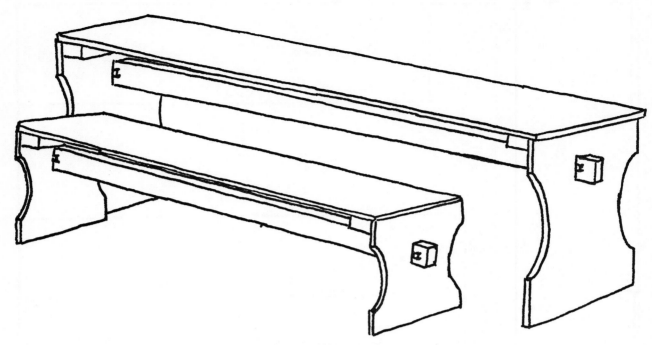

Fig. 9-9.   Trestle table and bench.

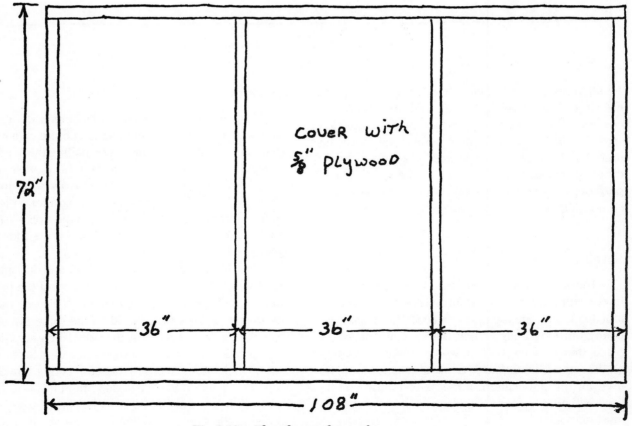

Cover with ⅝" Plywood

Fig. 9-10.   Floor frame of screenhouse.

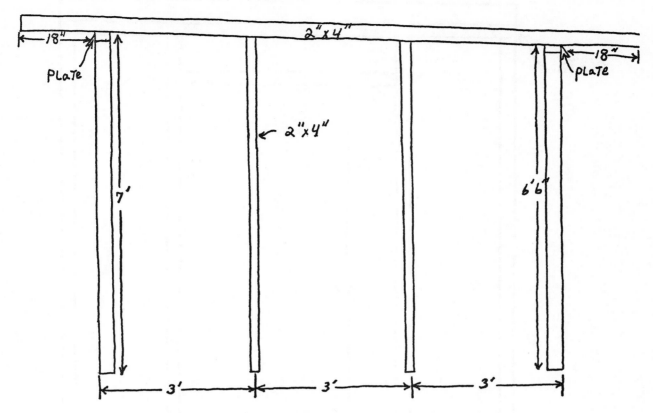

Fig. 9-11. Side frame of screenhouse.

deep as wide as the board. It is then fitted into the slot. Brass screws are used to join the joints together. The door is covered with screen that is held in place with screen bead.

Trestle Table and Benches

All the furniture to be used in the screenhouse can be handmade. It includes two benches and a table plus a double-decker bunk.

The table and benches, which are a matched set, are made in the trestle table design. Trestle tables, which are used for kitchen furniture, are made from very heavy stock. A card table doesn't need to be made this heavy. We can use ⅝-inch exterior plywood, sanded on one side for this tabletop. To prevent warping, 2 × 4 cleats are used. Start by sawing out a piece of exterior ¾-inch plywood 18 inches wide and 72 inches long. Sandpaper the corners. Next, cut three 18-inch pieces of 2 × 4 stock. Bevel each end of each piece. Fasten the 2 × 4's to the un-

derside of the plywood tabletop with 2-inch wood screws (see Fig. 9-15). The screws should be placed through the 2 × 4 into the tabletop. The legs are then fastened to the outside crosspieces.

The legs are also sawed from a piece of ¾-inch exterior plywood. Since they are visible from both sides, plywood that is good on both sides should be used. The initial sections are 15 inches wide and 27¼ inches high. They are made according to Figure 9-15. Shoes made of a double thickness of 2 × 4 are used at the floor to stabilize the table. The ends of the shoes are also beveled.

The 2 × 4's on either side of the pedestal legs are 25 inches long. They are tapered at the ends. The center joint or crossmember is a 2 × 6 long enough to reach from one end to the other of the pedestal legs plus a surplus of 4 inches on each end.

All joints are glued as well as fastened with screws. Two-inch wood screws are used in this application. When the table is done, it should be painted or stained with three or four coats of paint.

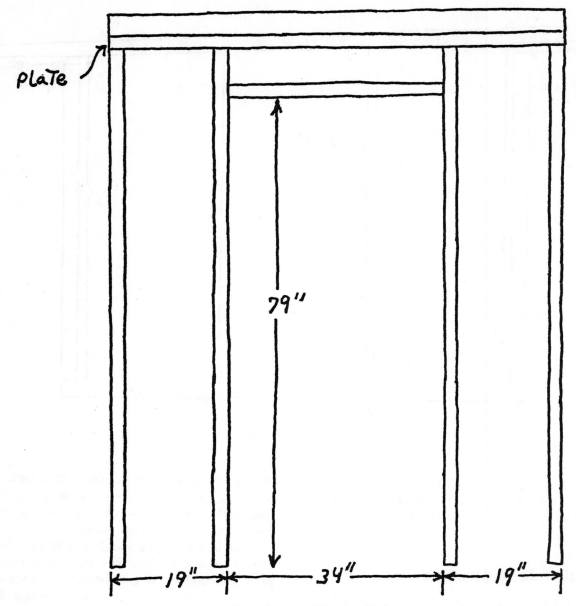

Fig. 9-12. Screenhouse end frame with door.

The benches are made almost like the table. The benches are 60 inches long, 18 inches high, and 12 inches wide. They too are made from ¾-inch plywood or the equivalent. Study Figure 9-16.

The crosspin between the pedestal legs is 2 × 4 instead of 2 × 6 as in the table. The benches should be finished like the table. The beauty of these pieces will be a result of how well you sand and finish them.

Double-Decker Bunk

Besides the table and benches a double-decker bunk with sleeping room for two to four people can be included in this screenhouse.

The bunk is made with a simple 2 × 4 frame. It is 52 inches high, 30 inches wide, and 6 feet, 6 inches long. The mattresses also are handmade from expanded foam covered with plastic material. The table is made small so it can be used at one end of the screenhouse or it can be used alongside the double bunk. If slightly crowded conditions are not objectionable, the table can be used at the end of the bunk.

Start building the bunk by cutting four knotfree 2 × 4's, 6 feet 6 inches long. Next, cut four 2 × 4's, 4 feet 4 inches long. Notch the 6-foot, 6-inch 2 × 4's so they can be fitted into the crosspieces. Each notch is 1⁵/₈ inches deep so a comparable notch in the crosspiece can be fitted into it and still maintain a level surface on the top (see Fig. 9-17). All pieces are attached to the legs with wood screws, through the legs and into the pieces. Further, three 2 × 4 crosspieces on which the mattresses rest are installed between the sides of the bunk. It is expected a sheet of plywood will be used under the mattresses if foam is used. If springs are used, then the springs should set above the plywood. An edge is formed by nailing a 1

× 4 to the sides of the bunk beds. For strictly sauna bathing and resting when the bather may have water on his body the plywood can be utilized without a mattress. A board bunk also is a fine place for resting on a hot day since it will usually feel cool.

FIREPLACE

Most outdoor type people need a collection of outdoor furniture which can be used when desired and of which no care whatsoever has to be taken. If it rains on it, it won't hurt it. If you forget and leave it out over the winter, very little damage will result. Just as important is that it be strong enough so the children can roughhouse on it and little harm will result. It also must have good looks and exude a rustic charm so the yard and buildings will be complemented by it.

One of the most useful of such items is a back-

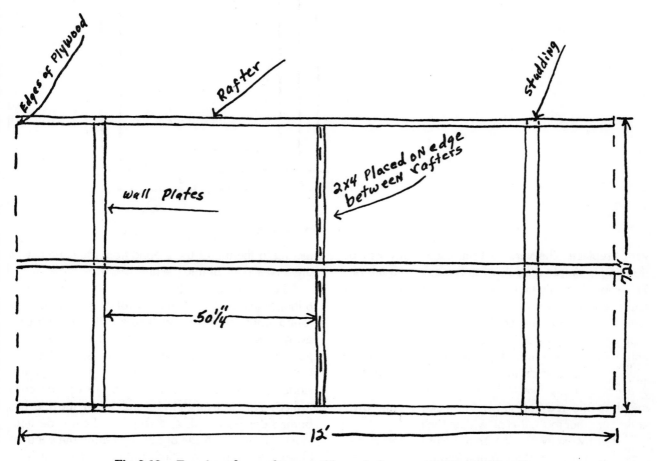

Fig. 9-13.   Top view of screenhouse roof frame before covering it with plywood.

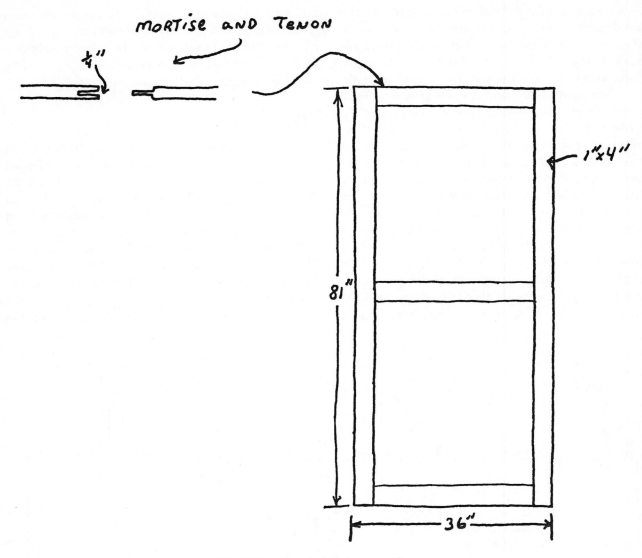

Fig. 9-14.   Screen door for screenhouse.

yard fireplace. The very best fireplace to my way of thinking is simply a circle of stones around a fire pit. This should ideally be made large enough so burgers and a roast can be cooked at one time. Such a pit also works very well for cooking a pot of beans or mulligan. The pit can also be used for cooking an entire meal at one time. Likewise, it can be used for smoking meat and drying food to preserve it. Also if you occasionally like to have some way of heating metal to the red-hot stage, for making your own knives or other tools, this pit will do the job.

Start by excavating an area 4 feet in diameter at least 1 foot deep at the top of a knoll if possible (see Fig. 9-18). First place cement tile in the bottom pro-

jecting to the top of the ground from the center. This tile serves to let air into the fire.

Further cover the bottom with flat stones at least 6 inches in diameter. At the side of the pit use colored round stones to form a wall so the fire doesn't drift away. The drain tile should be kept free of debris. A good plug for the draft tile is a round stone that will just about fit into it. By removing or installing these stones the amount of draft to the fire can be controlled. This is especially useful when the fire has burned down to coals and very high heat is desired. We even make a forge of ours by setting the vacuum cleaner to "blow" position and feeding the air into one end of the drain tile.

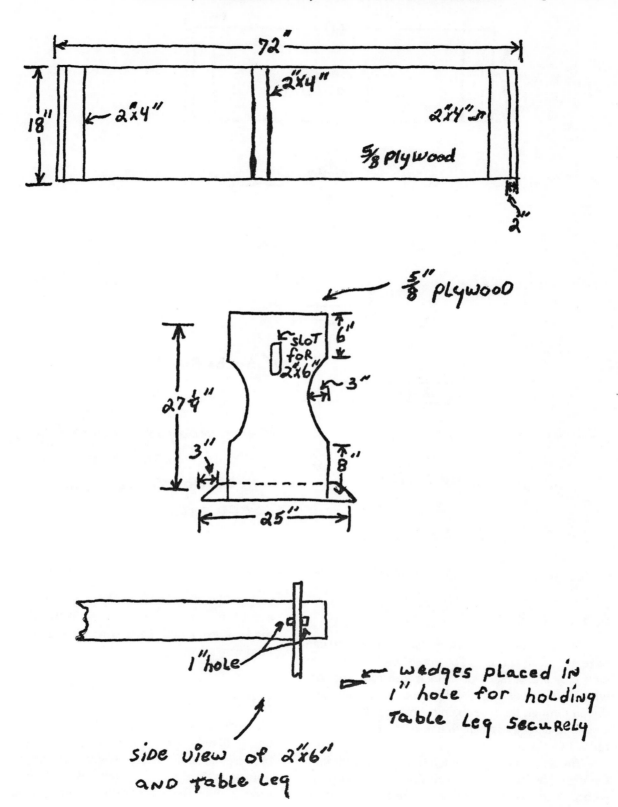

Fig. 9-15. Dimensions for trestle table.

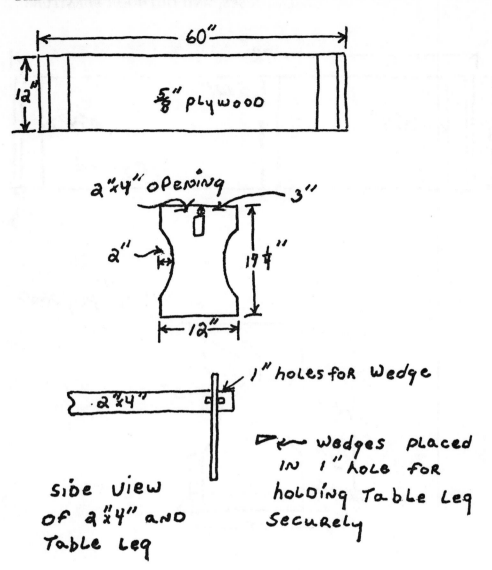

Fig. 9-16. Dimensions for trestle table bench.

Spit and Grill

The most useful accessories such a fireplace can have are a spit and grill. The entire spit is made from well pipe and a piece of ¼-inch iron rod. The rod, which is the actual spit, is sharpened at one end and driven through the meat being cooked. If it is not desirable to actually spit the meat, it can be tied to the rod with fireproof string.

The spit rotates in ¼-inch holes drilled through the pipe. The handle of the spit should be made from wood. A treelimb or similar contrivance could be used. It is held on with a cotter pin placed through the end of the steel rod (see Fig. 9-18). The grill can be made of expanded metal which is available at most hardware stores.

PICNIC TABLE

In conjunction with the fire pit a rugged picnic table will deliver many hours of enjoyment. The very best picnic tables to my way of thinking are made from logs. The top of the table is made from rough sawn lumber. Redwood is fine but fir, spruce, or pine will also do the job.

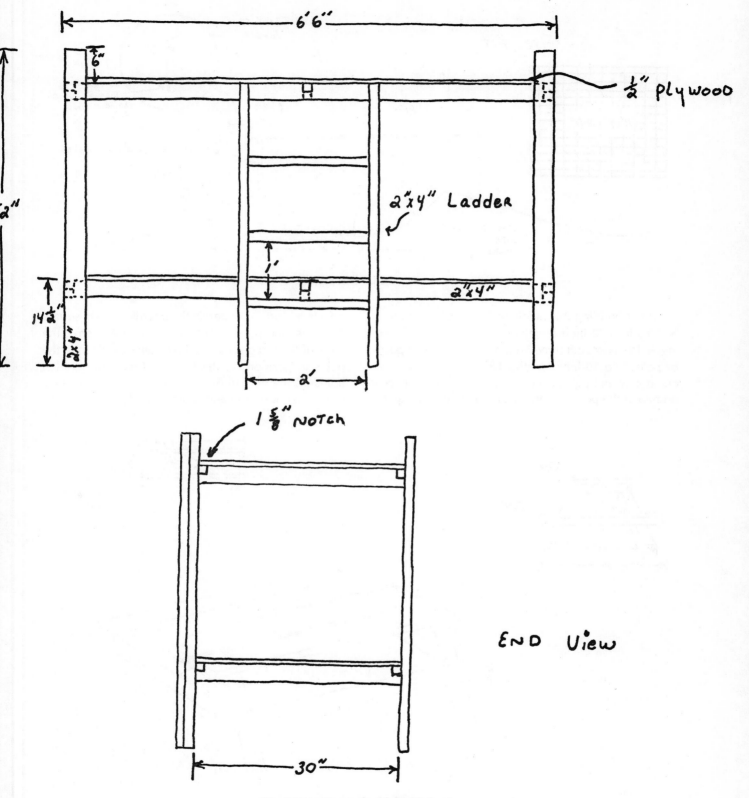

Fig. 9-17. **How to build double bunk.**

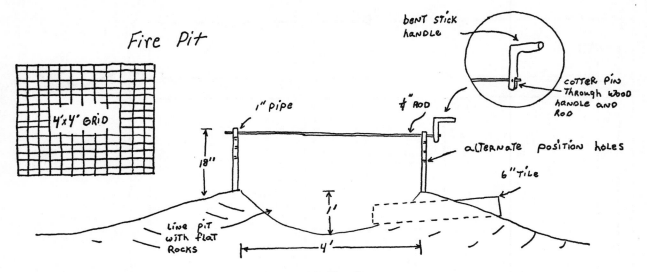

Fig. 9-18. Fireplace.

Start building this picnic table by cutting three 2 × 10's into 72-inch lengths for tabletop pieces (see Fig. 9-19). Next, cut three 5-inch diameter fence posts or poles into 30-inch lengths. Use the bench saw or chain saw and jig to rip two right-angle flat surfaces on two of the poles. Then lay out the three top pieces on a level surface, separate the boards ½ inch, square the ends, and position the two sawn poles across the ends of the top pieces. Clamp or temporarily nail the poles in position on the boards. Then drill ¼-inch holes all the way through the poles and top pieces. Make two holes in each top piece. Then place 6 ½-

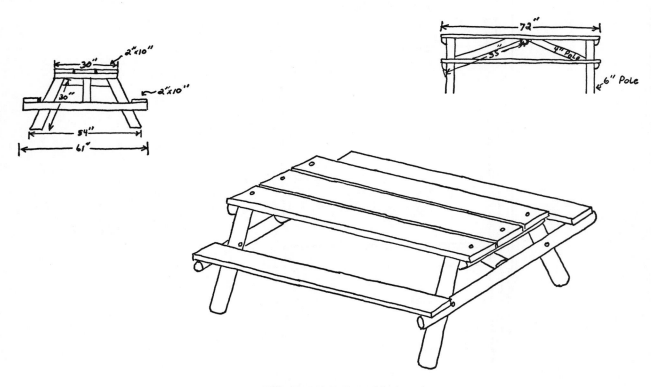

Fig. 9-19. Picnic table.

inch carriage bolts with the head at the top surface of the table through each hole. Place a nut and flat washer on each bolt and tighten them enough so the head is countersunk into the tabletop. Do this at all hole locations.

Now saw one flat surface on the remaining pole and place it at the centerline of the table. Use ¼-inch carriage head bolts to secure it across the top pieces also. This completes the top; the legs can be made up and attached next.

The legs are made from 30-inch-long, 8-inch diameter peeled poles or fence posts. Lay the poles out on a level surface parallel to each other and separated by about 54 inches. Next, cut a 4-inch pole, 55 inches long. Lay it across the legs 14 inches from the ends. Now drive a spike through the pole into each leg. Don't drive the spike down so the head is tight, since it must be removed. Next, pivot the other ends of the legs so they are separated by about 34 inches. This forms the correct angle for the notch which must be cut into the legs for the 4-inch crosspieces. Use a saw or scribe to make the marks for the notches. Pull the spikes out, remove the crosspiece, saw out the notches and replace the

crosspieces in the notches, drill the holes, and bolt them in place with 2¼ × 6-inch carriage head bolts at each joint.

Now attach the legs to the tabletop. Find the correct angle for this by laying a straight edge across the tops of the legs. Saw out this indicated line. Further notch out for the crosspiece at the tabletop. Drill two holes at each crosspiece and bolt them together with ¼-inch bolts. Next install a 35-inch-long diagonal brace of 4-inch diameter pole from the center of the crosspiece on the legs to the center of the crosspiece on the tabletop (see Fig. 9-19). Make up the legs for the other end the same way. Attach them and install a 2 × 10 plank between the crosspieces to form the bench. Finish the tabletop with at least four coats of a good clear preservative. If the heads of the bolts on the top of the table are objectionable, they can be countersunk and the hole filled with a section of wooden dowel. A 1-inch diameter countersunk hole and a 1-inch diameter dowel works very well for this.

Hardly anything can hurt this table and about the only care it will require is to be set on end in the winter so ice doesn't freeze to the top and spoil your winter outdoor picnics.

Chapter 10

RED BARN GARAGE

Many semicountry dwellers, with an acre or less to live on, keep animals, raise a garden, and house the automobile could find this building an answer to their prayers. It can be made as large as desired but one measuring 20 x 30 feet will cover most requirements.

All sorts of adaptations are inherent in this design. If there are no horses, both stalls can be used for goats and vice versa. When the room isn't being used to smoke meat or dry produce, it can be utilized as a workshop. The entire loft can be used for storing hay or other feed if desired. Approximately six tons of hay can be stored in this space if the hay is loose. The hay is fed to the animals through doors cut through the loft floor directly over the animal stalls. This loft can be used for a studio or an apartment also if modified to some extent.

It would ideally be set up with the animal stalls facing south in order to permit the doors to let in sunlight in the winter, which is important in animal hus-

bandry. If horses are kept in the stalls, window bars of wrought iron should be placed across the frame to prevent the animals from breaking the glass with their heads or from kicking the windows out. It wouldn't be too unusual for a 1200-pound horse to try to leap out the window either, even if it is barely large enough for him to get his head through. Horses seem to suffer from poor space-estimating ability.

Now, if after glancing at the plans and directions for this project, a homeowner decides it is too big a task for his untried abilities, let me reassure him. There is nothing about this building that can't be done by anyone who can hold a saw, hammer, or mason's trowel in his hands. Just proceed cautiously, doing each step in its proper place, and your building will take shape as well as if you were a carpenter with decades of experience. The hardest part is performing the initial steps. Perhaps if you do this in your spare time it will take a minimum of six months to complete. But, do you know any other way you can increase your estate $6000 to $7000 during leisure hours right in your backyard, in this amount of time and have fun doing it?

Don't forget also that this building can be used to house people. Many times a building of this type in conjunction with a conventional house will allow two dwellings to be built on the same lot and the farmstead will still exude the quaint charm of a rural setting.

It is also very adaptable to raising small animals such as hamsters, chinchillas, or beneficial insects such as praying mantis, and of course, can be used as a "factory" building for propagating "mousies," "waxworms," red worms, or any of the other insect larvae that are so much in demand for fishbait. Naturally, it can house that cottage industry that you want to start but don't have a place for.

LAYING OUT THE FOUNDATION

The first step in building this barn is to lay out the foundation. Generally, since it will be built on a small tract of land, the property lines can be used for guidelines. First establish a straight line at the property line or use the street or another building to orient the nearest side of the prospective foundation. An unorthodox, but effective way to locate a building if no other building, street, or marked property lines are close enough for reference is to use a magnetic compass for orientation. Do this by first deciding exactly where you want the southwest corner of the building to be located. Then sharpen a 4-foot length of 2 × 4 and drive it securely in the ground at this point, with the width oriented as much north and south as possible. Make sure it is straight up and down. Then lay a magnetic compass on the top of this stake at the center (see Fig. 10-1). Note where the north and south poles of the compass needle are and drive six-penny finishing nails into the stake at the indicated north and south poles of the compass. Remove the compass and put it away.

MATERIALS LIST FOR RED BARN/GARAGE

1. Approximately 100 8 × 8 × 16-inch concrete blocks
2. 20 2 × 4's for shoes and plates
3. 150 8-foot 2 × 4's for studding
4. 40 running feet of 2 × 4 material for window braces
5. 8 8-foot 2 × 6's for stall liners

6. 26 4 × 8 sheets of ⅝-inch exterior plywood for wall sheathing
7. 735 square feet of #15 builder's felt
8. 16 20-foot 2 × 8's for floor joists
9. 60 running feet of 2 × 8 for floor joist bridging
10. 19 4 × 8-foot sheets of ⅝- or ¾-inch plywood for the upstairs floor
11. 32 8-foot 2 × 6's for rafters
12. 32 10-foot 2 × 6's for rafters
13. 16 10-foot 1 × 8's for rafter bracing
14. 32 panel clips for rafter joints
15. 32 4 × 8-foot sheets of ⅝-inch plywood for roof sheathing
16. 725 square feet of roofing material
17. 108 running feet of 2 × 4 for gambrel studding
18. 160 running feet of batten boards
19. 132 running feet of fascia
20. 2 4 × 8-foot dutch doors
21. 2 conventional entry doors
22. Loft door
23. 8-foot garage door
24. 4 windows (size to be determined by builder)
25. Anchor bolts and nails

Now the six-penny nails can be used as a north-and-south sighting plane. Proceed to lay out the foundation by measuring north 30 feet from the original stake. Use a thin pole or a 2 × 4 edgewise for a surveyor's pole and adjust it until it is directly north of the original stake as indicated by sighting across the nails. A level should be used with the surveyor's pole to make sure the top and bottom are perfectly vertical. When this second position is located, drive a stake at this point. Now the north and south line of the foundation is located and you can proceed to lay out the rest of the foundation from this north-south line according to the suggestions on laying a greenhouse foundation in chapter 2.

SQUARING THE FOUNDATION

A method often used by carpenters for squaring a foundation is called the *6-8-10* method. After the preliminary corners are found and strings stretched all around the foundation, select a light, easy-to-handle pole and adjust its length to exactly 10 feet. Then put a marker on the front wall string exactly 6 feet from either corner. This marker can be a short length of tape or simply a chalk mark. Further, put a marker on the side wall string exactly 8 feet from the same

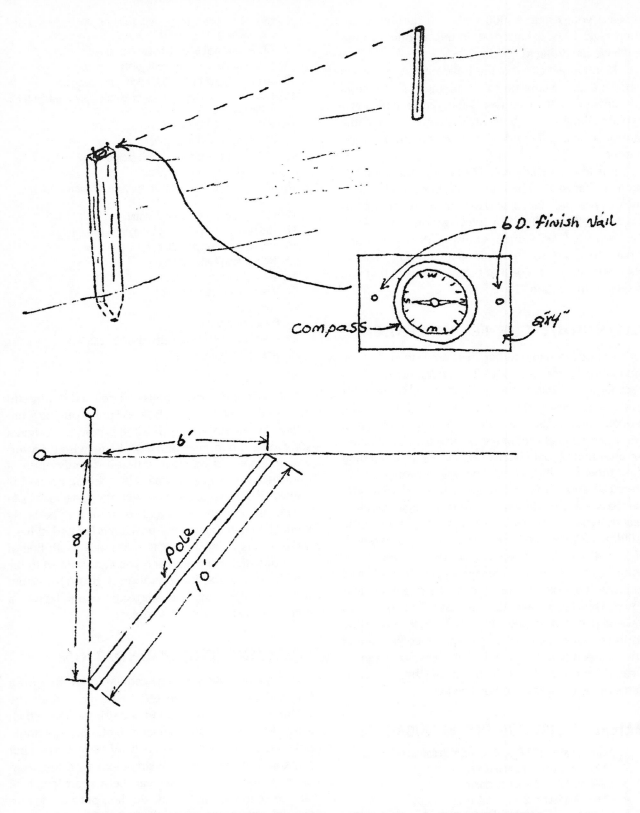

Fig. 10-1. *Top:* Locating a building line. *Bottom:* The 6-8-10 method of squaring a corner.

corner. Now the 10-foot pole should reach exactly from the 8-foot mark on the side wall string to the 6-foot mark on the front wall string if it is held level between the two marks (see Fig. 10-1). If it does not, then adjust the *side wall* string until it does. With this corner squared, repeat the procedure at either back wall corner and the entire foundation will be "squared" and "true."

FINDING THE GRADE LEVEL

The next step is to find the grade level. First, drive a stake in the ground at what looks like the highest level. Further, drive a nail in the stake and tie a string to the nail. Place a line level on the string and check all points which will be inside the foundation. If the slope of land is more than 4 inches some earth has to be removed or the foundation wall has to be built two blocks instead of one block above the natural soil line.

LAYING OUT THE EXCAVATION LINES

Now the excavation lines can be laid out. Generally, this is done with batten boards. Three feet outside each foundation corner erect a fence of scrap 1 × 4 boards with the horizontal rail 18 inches high. The "fence" should extend past each corner 3 feet. Do this at all four corners. Now extend the original foundation strings to the batten boards so they cross directly over the original corner stakes. Saw slits in the batten boards to level the lines and weigh each line with a brick so the line will be tight but portable (see Fig.10-2). Further, measure 30 inches outside the original lines and make another slit in each batten board. This second slit indicates the excavation line which the equipment operator will use to dig the trench for the cement block foundation.

Now either the building line or the excavation line can be quickly found by transferring the brick-

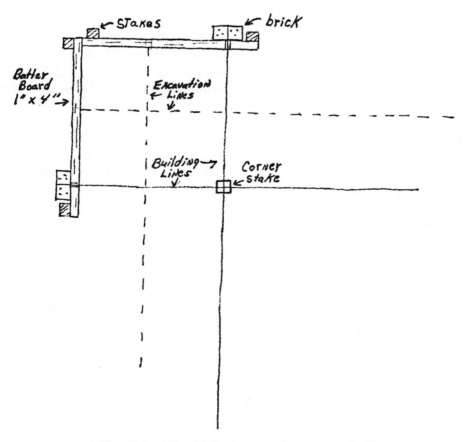

Fig. 10-2. Using batten boards to find excavation lines.

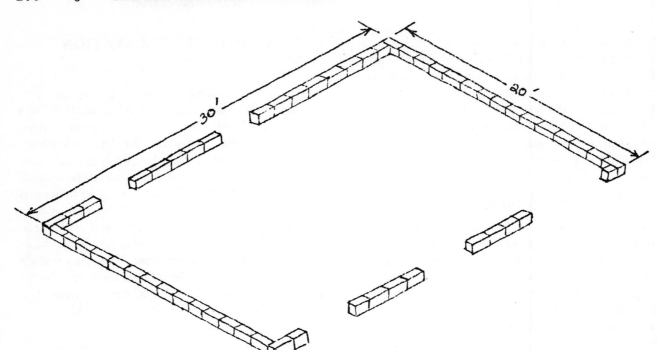

Fig. 10-3. Top row of foundation blocks.

weighted string from one slit to the other. Of course, if you dig it yourself, this will also determine where you wield the pick and shovel.

Now, with the excavation lines laid out, the depth of the excavation will be the next consideration. Generally this should be below the frost line—4 feet below grade in the North, at least 3 feet in the South—but in all instances it should be dug down to firm mineral soil. Place footings under the wall. They should be at least 3 inches thick and 16 inches wide. Consequently, the excavation will have to be widened at the bottom to accommodate them. After the footings have set about three days, the block laying can commence.

BLOCK LAYING

The methods of block laying will be the same as for the full-size conventional greenhouse in chapter **2.** We suggest the building be built at least one course of blocks or 8 inches above ground level to prevent water seepage problems. The blocks will have to be left out of the door openings to permit entry at the ground level (see Fig. 10-3). Don't forget to place the anchor bolts or anchor clips in this top

row of blocks. If anchor bolts are used, plug the opening in the cement block with paper or a similar material at the bottom of the top course. Then fill the opening with concrete to the top of the course. Set the bolt in position with a washer at the head. The washer, of course, provides greater anchorage than the head of the bolt used alone. The bolt should extend 4 inches above the block. Locate the bolts by snapping a chalk line 3½ inches from the inside edge of the 8-inch blocks. Imbed the bolts, spaced 2 feet from each corner, 4 feet apart on straight walls. Also the door frames can be anchored to the blocks with bolts or clips, with clips being more expedient. Place the clip at the joint under the top course of blocks.

After the perimeter walls are in place, the interior partition walls can be laid. Six-inch instead of 8-inch blocks can be used for the interior walls if desired. Generally, the interior walls should be buried at least 8 inches, extended 4 inches aboveground. Make sure the bottom course of blocks rests on mineral soil, however. Poured partition footings can be used also. They should be at least 4 inches wide and 4 inches aboveground, and buried at least 8 inches or to a depth where they can rest on undisturbed mineral soil. Imbed anchor bolts in the interior footings also.

Use a concrete mix compounded of one part cement, three parts sand, and five parts 3/4-inch gravel for poured walls.

FLOOR

At this time the type of floor that will be used can be considered. If the building codes specify concrete, it will be easier to install it now than when the building is put up. Actually, the floor in the animal stalls should be left as undisturbed dirt if possible. The floor in the automobile section can be traffic-compacted crushed gravel and the floor in the other two sections can be made from crushed gravel or treated lumber.

In no case should horses or other animals be required to stand for long hours on bare concrete. The floor should be covered with planks or with a thick layer of straw if a concrete floor must be used.

In suburban localities, local building codes often specify that the stable floor be drained into a dry well. Probably the local building inspector will notify you of this when you go to get the building permit. He may have plans for a dry well also. He might want to know what you plan on doing with the manure. In these days of organic gardening it usually can be sold to eager vegetable and flower raisers, and some of the cost of the feed can be returned. Also, any nearby farm will be glad to get it, especially if you deliver it to the fields. Naturally, if you build on sufficient acreage, none of these problems will be encountered.

THE ART OF DEALING WITH BUILDING INSPECTORS

Dealing with building inspectors is more of an art than a science. They vary from very conscientious, experienced men to outright con artists. Some even try to stop private citizens from doing their own building, or at least derive as much income from them as possible if they do. Here are two ways a homeowner can force fair treatment from a hostile, opportunist-type building inspector: (1) Very earnestly call him up two to three times a day and ask questions about the building. After this goes on for about three weeks, the inspector is likely to shout, "Don't call me again. I'm not your contractor." (2) Make a great show of writing down or recording every word the inspector says. He will probably become very close-mouthed when faced with this situation.

The most self-serving way to deal with the building inspector is to use his knowledge to improve your structure, but don't allow him to bully you out of building or into hiring any of his contractor friends.

WALL FRAMING

Anyway, after the flooring and partition footings are in place, the framing can commence. Essentially, wall framing consists of the bottom horizontal members called shoes, top horizontal members called plates, and the vertical members between the top and bottom called the studding. Generally, all of these members are made of 2 × 4 lumber. It is recommended in this building that both the shoes and plates be doubled.

Make arrangements to buy, rent, or borrow an electric hand held or table saw. Even if you buy one, it will pay for itself before this project is half-done.

You can start by framing in the west end wall. Select two 10-foot 2 × 4's as knotfree as possible. Saw each end of each piece if necessary to square the ends, butt them together, and make sure the overall length of the two is near 20 feet. Next square the ends of a third 10-foot 2 × 4 and nail it so the center, or 5-foot mark, falls directly on the joint of the previous two 2 × 4's. Use ten-penny nails to nail the 2 × 4's together.

Complete the assembly by nailing two 5-foot lengths of 2 × 4's at each end. This completes the shoe, spliced together for maximum strength. Next, carefully measure and saw the shoe to a length of 19 feet 5 inches. This is necessary because the side walls are cut a full 30 feet long and the thickness of each side wall must be included in the width since building measurements are taken across the outside. Before the studs are nailed to the shoe, lay the shoe alongside the anchor bolts in the foundation wall and mark the location of each bolt hole on the plate. Drill the holes 3/4 inch in diameter for ease of assembly.

The next step is to fabricate the plate. This is done identically to the shoe except that the top board of the double 2 × 4's is left a full 20 feet long to tie the end wall into the side wall (see Fig. 10-4). In all cases the splices should be made so they fall directly over the end of a stud, and the plate and shoe splices

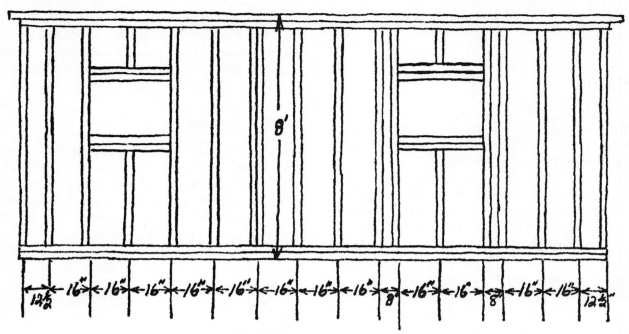

Fig. 10-4.　West wall frame.

should not fall on opposite ends of the same stud. If good straight 20-foot 2 × 4's are available, they should be used to avoid the splices.

After the shoe and plates are made up, lay the plate on the shoe with the 19-foot, 5-inch section of

the plate flush with the ends of the shoe. Use a pencil to mark the location of each stud on both the plate and shoe. This will eliminate measuring each one after they are installed. The studs are cut 7 feet 6 inches long except for the short lengths over and

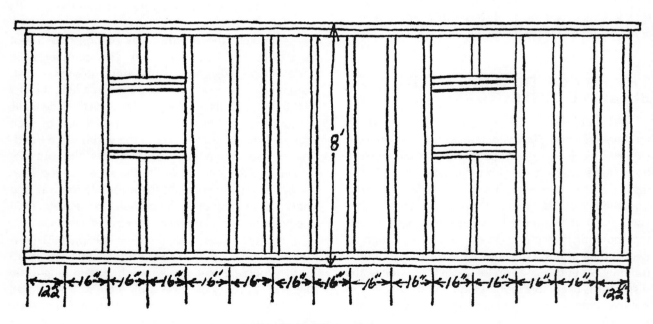

Fig. 10-5.　East wall frame.

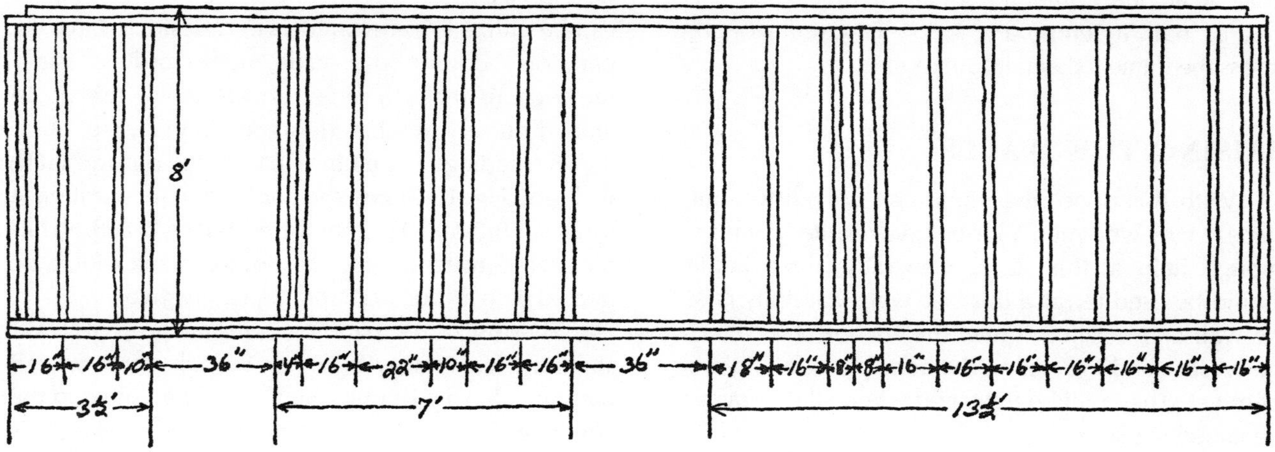

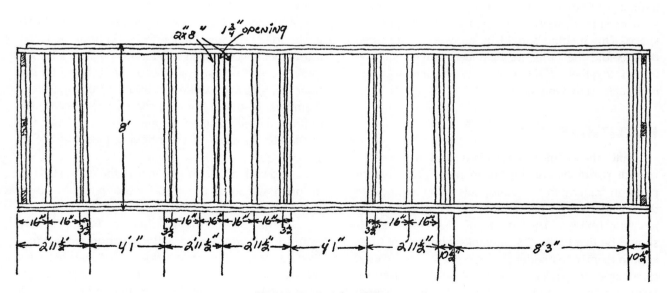

Fig. 10-6. North wall frame.

under the window frames. One hundred fifty 8-foot studs will be required in all. Study Fig. 10-9.

Notice at the top and bottom of each window frame a double 2 × 4 is required to prevent warping. Before the west wall is made up, the rough openings for the windows should be checked with the suppliers in your locality to see if they have windows that will fit. Then make the openings according to the available windows. Fail to do that and you may have to do what the author did and settle for a window much smaller than he really wanted during one building project, simply because no window was available that would fit the rough opening called for in the plans.

The studs should be fastened to the plate and shoe by driving sixteen-penny nails through the plate and shoe and into the end of the stud or by toenailing two eight-penny nails on either side of the stud into the plate and shoe.

After the wall is completely framed, use a 50-foot tape to measure diagonally from both top corners to both bottom corners to see if it is square. If it isn't,

Fig. 10-7. South wall frame.

strike it at one corner to square it up. When it is square, nail a temporary 1 × 4 brace diagonally across the frame to keep it square.

RAISING THE WALLS

With this done, the frame can be hoisted into place. Many workmen injuries have resulted from not enough help at this stage. Generally, three adults should be used to raise this size wall. Carefully raise the wall and position it over the anchor bolts. Place the washer and the nut on the bolts but don't tighten them yet. They will be tightened when all the frames are nailed in place.

When the frame is in place, make sure it doesn't come crashing down again by bracing it in at least two places by extending 2 × 4's from the frame to the ground. At least the first and second frames being raised into position should be made plumb. Do this by driving an eight-penny nail into the end of the plate and hanging a plumb bob from it. Check the frames for alignment by stretching a chalk line from one end of a frame to the other at the side of the plate. If it is found to be either out of plumb or crooked, straighten the offending member by pounding on a piece of scrap wood placed against it.

Continue to fabricate and raise each wall into position. If, as frequently happens, a wall must be shimmed between the shoe and the concrete to level the frame and space is left under the shoe as a result, a mixture of one part portland cement and three parts sand must be flushed under the shoe to close up this space. This mixture should be allowed to dry three days before the bolts are tightened down. Nail the frames together at the corners with sixteen-penny nails, at least two in each location.

PARTITIONS

After the exterior framing is up and in place, the interior partitions can be put in place. The interior partition framing in this building is made identical to the outside walls. This simplifies construction for the amateur builder. The inside of the animal stalls should be lined with 2 × 8 planks if horses are to be kept in them, or a divider of 2 × 8 planks can be used to separate the stalls. If the stalls won't be used for horses, a solid partition of 2 × 4 studding covered

with plywood or a grillwork to promote ventilation can be utilized. Some builders may desire to have the partition between the stalls made so it is easily removed to make a larger enclosure for use as an animal nursery or a breeding stall. In that case, place 2 × 8 studdings 1¾ inches apart in the outside south wall (see Fig. 10-7) and the partition wall (partition A in Figs. 10-8 and 10-9) and bolt 8-foot 2 × 8 planks on edge between them. The bottom plank should be treated with a good waterproof preservative.

The rest of the partitions are made up of 2 × 4 studdings covered with ½- or ¾-inch plywood. All partitions have a double-thickness shoe bolted to the concrete divider walls, but a double-thickness plate is optional in the interior walls. When the partitions are complete, the floor joists can be installed.

FLOOR JOISTS

Most codes call for 2 × 10's to be used with the 10-foot span between the walls in this building. Generally, 20-foot 2 × 10's are available but if they are not use 12-foot 2 × 10's and splice them over the plates. Use only knotfree, straight boards, but if a crowned or crooked board must be used, position the crown up since it will straighten in time. Nail the joists together at the splices with eight-penny nails and into the plate with sixteen-penny nails. Also notice that the first floor joist is nailed 3½ inches from the gable end. Each succeeding joist is nailed on 2-foot centers (see Fig. 10-10).

After all the floor joists are in place, cut and nail in the joist bridging. Generally, solid bridging is the most advantageous to install. Simply cut 2 × 10 stock into 21-inch pieces. Nail the bridging between the floor joists at the maximum distance from the cross members. Also, the bridging should be staggered for optimum utility and to permit nailing through the joists into the ends of the bridging pieces (see Fig. 10-10).

Also nail in extra bridging over the horse stalls so an opening can be provided for dropping hay down to the horses from the mow.

By this stage in the building, you have probably noticed that you are not tall enough to work effectively from the ground. A ladder, of course, provides the easiest-to-obtain platform. If two sturdy ladders are available, they can be used in conjunction with

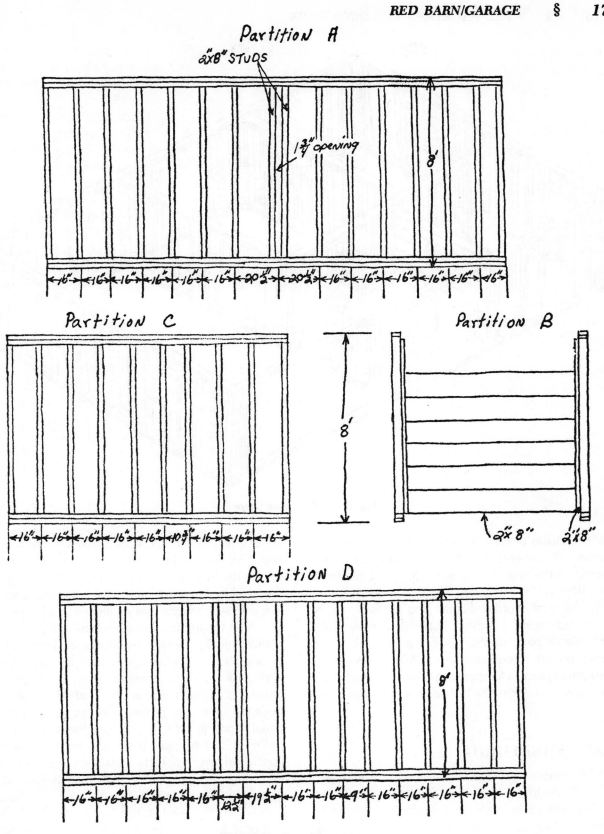

Fig. 10-8. Interior partitions.

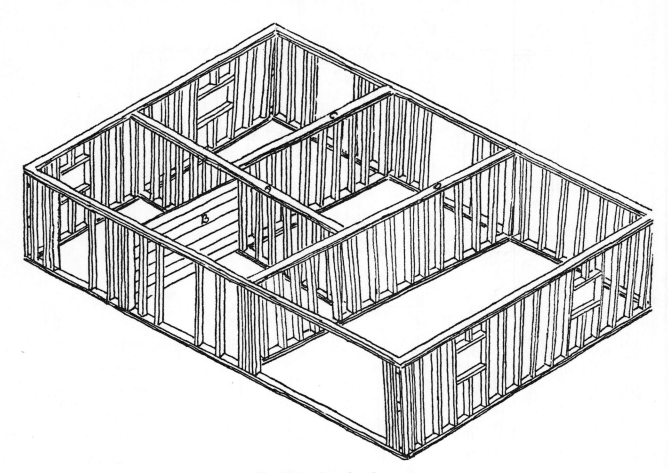

Fig. 10-9.   Complete frame.

ladder jacks to provide a working platform. The old standby of carpenters is the wooden scaffolding, erected on the spot. This makes a fine, safe platform but it does cost extra money and time to build. Fortunately, in most populated areas pipe scaffolding can be rented from building material suppliers or hardware or paint stores. You can also use 2 × 10 planks placed across sawhorses. After a few floor joists are in place, a temporary platform in the form of a few plywood sheets can be laid on the joists to work from.

WALL SHEATHING

The next logical step in construction is to apply the wall sheathing. Building codes often call for building felt to be used under exterior sheathing. Number 15 or, as it is commonly called, 15-pound felt is used. Apply it horizontally and overlap the edges 2 inches. Further, it should overlap the foundation 2 inches. It will take three runs of 36-inch-wide felt for the side walls of this building. A staple gun is the ideal tool to use to fasten the felt to the studding. Place a 1/2-inch staple every 4 inches on all edges.

After the felt is applied, the sheathing can commence. Ideally, 5/8-inch exterior grade plywood will be used for this, although 1/2-inch or even 3/8-inch may pass the code in some areas. Every plywood panel should butt together on a stud. This building is designed to accept pieces of plywood. Be sure the panel is positioned so it is 1/2 inch lower than the upper edge of the upper plate. Use eight-penny nails positioned every 6 inches apart on each stud and each edge. Each end panel sheathing should extend past the corner 5/8 inch to cover the ends of the side panels (see Fig. 10-11). On the 30-foot walls there will be 1 foot of waste on each panel which can be utilized in building the horse mangers. Cut out the

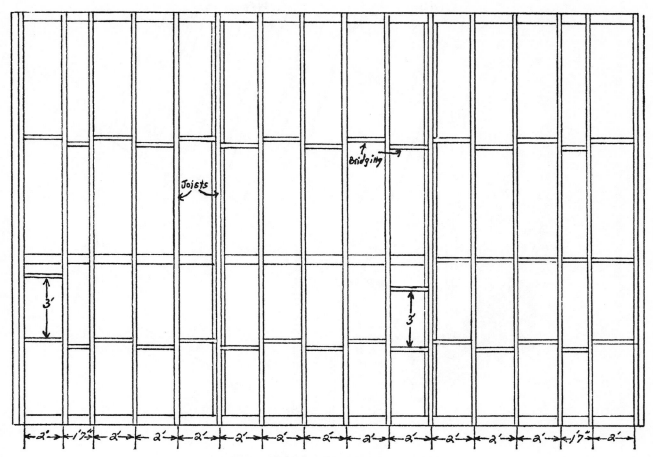

Fig. 10-10. Floor joists.

window openings after the panels have been installed.

HAYLOFT FLOORING

After the side wall panels are installed, the ceiling or hayloft flooring can be installed. Use ¾-inch exterior plywood if available. Start at the center by snapping a chalk line from one end wall to the other on the 10-foot line directly in the center. Align the first row of plywood panels carefully on this line and the remaining rows will be straight. Nail to the floor joists every 8 inches with eight-penny nails. Do not place the outer panels at this time since they need to be out of the way for installing the rafters. Always butt panels together over a floor joist. If you "miss" a joist, it will be necessary to install an additional bridging to brace the butt ends of the paneling.

MAKING AND INSTALLING THE RAFTERS

When this is all done, it will provide a platform for fabricating and installing the rafters. If you don't wish to make rafters, your local lumberyard will no doubt do it for you. However, you can do it yourself handily enough by following the simple procedures outlined here. First, select 32 16-foot 2 × 6's. Make sure they are straight and knotfree. Fir or pine will have sufficient structural strength. Cut a 9-foot length and a 6-foot, 6-inch length from one 16-foot 2 × 6. These two boards will form two angles of one side of the gambrel rafters.

Now take the boards, an 8-foot straight edge, a 20-foot steel tape, and a carpenter's square and climb up to the loft floor. Make a line (*A* in Fig. 10-12) 7 feet, 9 inches long on the loft floor, lengthwise of the building about 3 feet from the edge. Use the carpenter's square to find the corner and project a

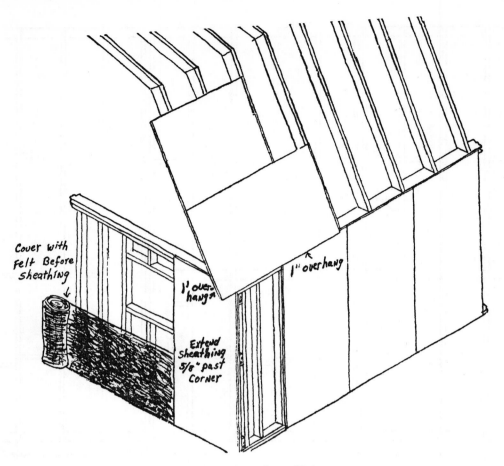

Cover with Felt Before Sheathing

1' over-hang

1" overhang

Extend Sheathing 5/8" past Corner

Fig. 10-11. **Sheathing the stable/garage.**

second line (*B*) 4 feet, 3 inches long at right angles to the first line towards the center of the building. Now try the 9-foot length of 2 × 6 for connecting the ends of line *A* and *B* as shown (see Fig. 10-12). If your measurements are correct, this should just fit. If it does, project line *C* 3 feet in the same direction as line *A*. At the 3-foot mark of line *C*, project the right-angle line *D* 5 feet, 9 inches long. Now try the 6-foot, 6-inch 2 × 6 between the ends of lines *C* and *D*. It should just fit. If it doesn't, go back and check the measurements. If it does, proceed to find the angle cuts for each of the two boards. To find these angle cuts project lines *D* and lines *A* until they intersect. Then draw a line (*E*) from the intersection of lines *A* and *D* approximately 6 inches past the junctions of lines *C* and *B* (see Fig. 10-12). Then lay the previously cut 2 × 6's, one at a time, in their proper position and mark both edges where line *E* falls. Connect the marks with a square and saw them off. The angles for

the ridgeboard and the plate can be found with the square by simply extending a line at right angles to lines *A* and *D* in both cases. Study Figure 10-12.

Notice that a 1 × 8, 5 feet long, is used for a brace across the joint on both sides of the rafter, on all except the end rafters (see Fig. 10-12). Also a panel clip should be used at the joint, or a 1 × 8 × 12-inch wooden gusset plate can be made up and nailed across the joint. Nail 1 × 6 collar beams across the rafters after they are in position.

It will be necessary to tack one rafter together temporarily and check it with a temporary ridgeboard to see if it fits. If it doesn't, make any adjustments necessary and then disassemble the rafters and use each board as a pattern for making up the rest of the rafter assemblies. Notice that they are spaced on 2-foot centers and thirty-two will be required in all (see Fig. 10-13). Always use the pattern pieces for making all of the others to decrease the chance for errors.

When the time comes for installing the rafters, temporarily bolt the lower end of the first rafter to the floor joist using one $1/2 \times 3$-inch bolt. Then tack it to the ridgeboard with an eight-penny nail. If both ends line up well, drill the second hole in the rafter and floor joist and install the second bolt. Then toenail the top end of the rafter to the ridgeboard with two ten-penny nails on each side of the rafter. The two end rafters have to have a spacer block nailed to them to bring the rafters flush with the inside joist.

The ridgeboard has to be held in position with a fixture until several of the rafters are in place. Make up this fixture by tacking a 1×6 to a stand made by nailing a 4-foot 2×4 to the loft floor (see Fig. 10-13). Use duplex nails or leave the heads of the nails slightly raised for easy removal.

After all the rafters are nailed in position, plumb them up and nail a scrap 1×6 brace across them to keep them true while the roof sheathing is applied.

Notice that the brace is underneath, not on top of the rafters. The ends of the floor joists extend beyond the angle of the rafters and they will have to be sawed off to the rafter angle.

LOFT SHEATHING

The next step after the rafters are in position is to nail the rest of the loft sheathing into place. It has to be slotted to fit past the rafters. Mark for these slots by holding the panel against the rafter and projecting the dimensions on the panel with a square and pencil. Use a saber saw to saw out the slots. With this done, the end gable studs can be installed.

FITTING THE GABLE STUDS

Fitting the gable studs looks more complicated than it is. Simply stand a 2×4 in the proper position

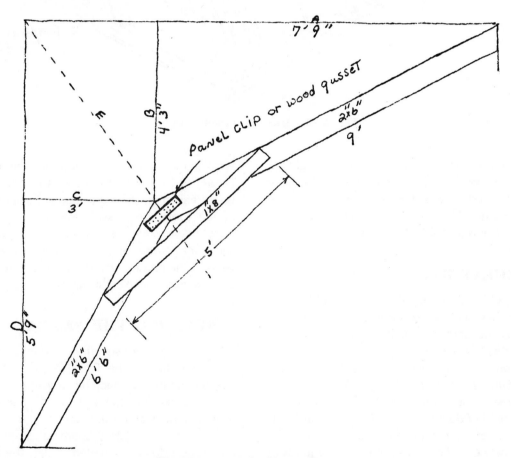

Fig. 10-12.   Gambrel barn rafter.

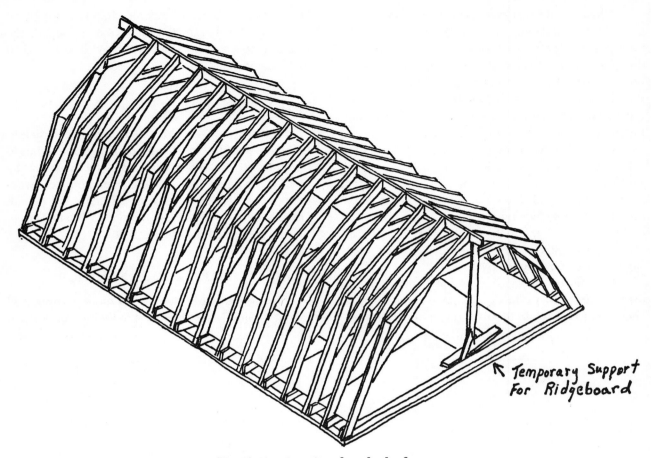

← Temporary Support For Ridgeboard

Fig. 10-13.   Location of gambrel rafters.

on the plate, use a level to make sure it is straight up, and draw a pencil line where the gable rafter intersects the stud. Notice that the west end has a loft door opening (see Fig. 10-14). This takes an opening liner of 2 × 4 plus a double 2 × 6 header.

ROOF SHEATHING

Now, with this done, the roof sheathing can commence. As with most roofing work, the first course must be started straight with the building. Generally this is done by snapping a chalk line to show the proper position for the upper edge of the plywood. Since a 1-inch overhang is desirable, snap the chalk line 47 inches up from the end of the rafters. A good method of doing this is to start six-penny nails on either end of the rafters and stretch and snap the chalk line between them. Do not place the joints of the panels parallel with each other since it could

weaken the roof. One way to install them is shown in Figure 10-11. When the peak is reached, the ridgeboard has to be shaved to prevent a space at the top of the roof. Use a skill saw set at an angle or a drawknife to slant the ridgeboard to the angle of the sheathing. Notice that the panels project 12 inches beyond the end rafters. The panels should be nailed every 8 inches with eight-penny nails.

INSTALLING THE FASCIA

When the roof sheathing is complete, the fascia can be installed (see Fig. 10-15). Notice that the 1 × 6 end fascia is made like a gambrel rafter. This is nailed to the ridgeboard at the top and through the roof sheathing along the slant of the roof. The side fascia consists of 1 × 6 boards also and it is nailed under the side eaves, butted against the roof sheathing. When this is done, the roofing can be applied.

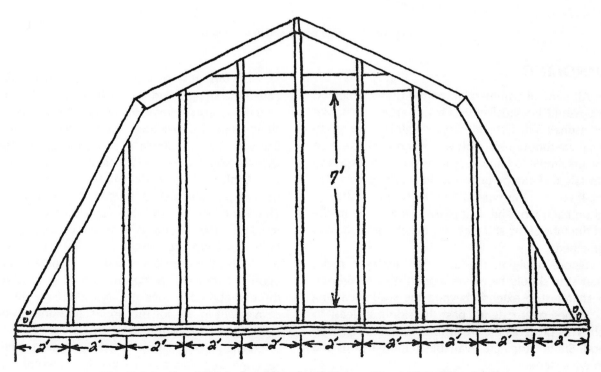

Fig. 10-14. *Top:* West end studding detail. *Bottom:* East end studding detail.

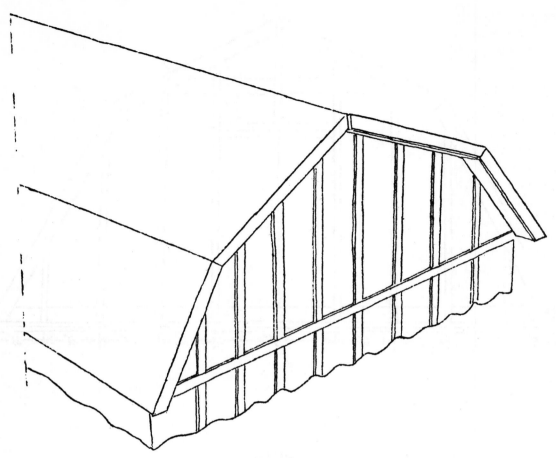

Fig. 10-15. Roof fascia and batten boards.

SHINGLING

All sorts of commercial roofing is available and most would be satisfactory for this application. We used asphalt 3-in-1 shingles that weigh 300 pounds to the square for our roof. This was a good choice since they are simple to handle, require very little expertise to install, and can be put on while working from ladders. If you use this type shingle, be sure yours have a seal on each tab. The heat of the sun will eventually seal the tabs of the shingles to prevent the wind from lifting them.

Before applying the asphalt shingles a metal starter strip should be nailed along the eave. Be sure the starter strip extends beyond the wall sheathing 1¼ inches so it extends past the siding at least ½ inch. This prevents the water from running down the wood. Starter strips are available at lumberyards and hardware stores.

The first row of asphalt shingles at the roof is called the shingle starter strip. It is a standard shingle nailed on upside down so the notches between the shingle tabs are extended towards the peak and the flat edge of the shingle is along the eave. Use six large-headed roofing nails per strip.

Apply the succeeding rows of shingles in the normal position, overlapping each about 5 inches. The notch between the shingles is used as a guide in regulating the amount of overlap. Place the nails 1 inch back and 1½ inches on each side of the notches.

When the peak is reached, use a row of shingles applied parallel with the ends of the roof. Overlap this row about 5 inches also. A metal ridge roll also is convenient to use for this application and gives the roof a finished appearance.

When working from a ladder, some method of securing the ladder so it doesn't slip is recommended.

Many roofers use a ½-inch nylon rope tied from the top of the ladder over the roof ridge and secured on the other side to a door, window, or wall anchor.

DOOR AND WINDOW FRAMES

After the roofing is put on, the door and window frames can be made up and installed. As mentioned before, be sure to check what is locally available in window sizes before the openings are left in the studs. Window frames are made according to a definite standard procedure. First, cut two 1 × 4's to the size required to line the vertical sides of the rough opening. Then carefully measure the horizontal length of the rough opening, deduct the thickness of the two vertical 1 × 4's and saw out a 1 × 6 to fit between them. These will form the top and sides of the frame. The windowsill is made from a 2 × 6. It is sawed to be placed flush with the back edge of the vertical sides of the frame and notched to extend past the sides of the vertical frame for 3½ inches on each side. This is to cover 4-inch window casings. This window frame should be installed flush with the inside edge of the studs. If it is desired to use 2 × 8 stall liners in the animal stalls, be sure to use 1 × 6's for the vertical and horizontal members. Further, use a 2 × 8 for the windowsill.

Once the window frame is made up, place it in the rough opening and use shingles to square it up. When it is square, nail it to the studding with sixteen-penny nails. Make sure it is flush with the outside sheathing before nailing. When the window is in position, nail 1 × 4 casing all around the perimeter. A drip cap made from a 1 × 2 pine strip can be installed as an optional installation.

Many times windows are made up by the manufacturer complete, so the window frame and sash can just be installed in the rough opening. All that has to be done then is to nail the window trim in place.

The door frames of 1 × 6 boards on the north wall are nailed inside the studding which forms the 36-inch rough openings for the doors (see Fig. 10-16). The door frame is squared up inside the rough opening with shingle wedges and then nailed securely in place with ten-penny nails. A drip cap and 1 × 4 casing is also used with these doors. The two dutch doors

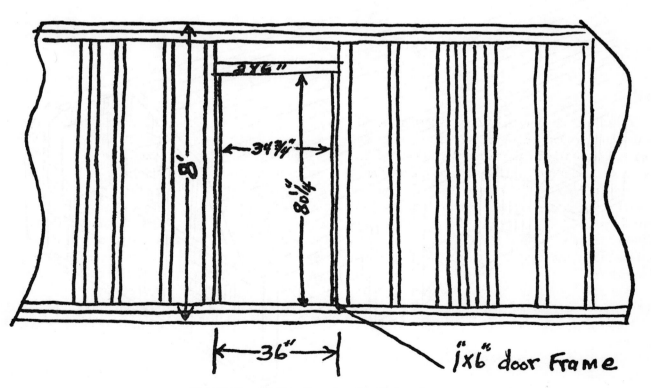

Fig. 10-16.   Section of north wall with door frame.

on the south w∷ are also framed in this way, but 2 × 8 headers must be used for this length of opening.

The garage door is framed by lining the inside of the 8-foot opening with 2 × 8 planks. Generally adequate instructions come with garage doors so they can be hung securely. Try looking at a neighbor's garage door installation if you feel insecure about installing one. For this type of installation an overhead track-type door should be satisfactory, although tilting, folding, and sliding doors are also good choices. Generally, whatever is available within the budget is a good choice since any of these types of doors are satisfactory if installed correctly and maintained well.

The dutch doors used on the south wall probably must be handmade since purchasing custom-made doors like this is nigh impossible. Perhaps the easiest and fastest way to make a dutch door is to cut out a piece of ¾-inch exterior grade plywood to the desired shape. Line one side with tongue-and-groove 2 × 6's (see Fig. 10-18) by placing number 10 flatheaded wood screws through the plywood into the liners. When it is assembled, saw the door apart and then place a cleat across the top section so that it overlaps the bottom. This can be a 1 × 4 fastened to the top section with wood screws or nails. For a "professional" look countersink all nail holes and fill the holes with putty. Personally, I don't find the head of a nail offensive; in fact, I find it reassuring to see the correctly placed nail or screw heads giving evidence of the fasteners buried in the wood doing their darndest to keep my building together. Naturally, the nails should be evenly spaced, however, if they aren't covered.

When the door is done, proper hardware should be secured for it. Many carpenters weigh a door on a

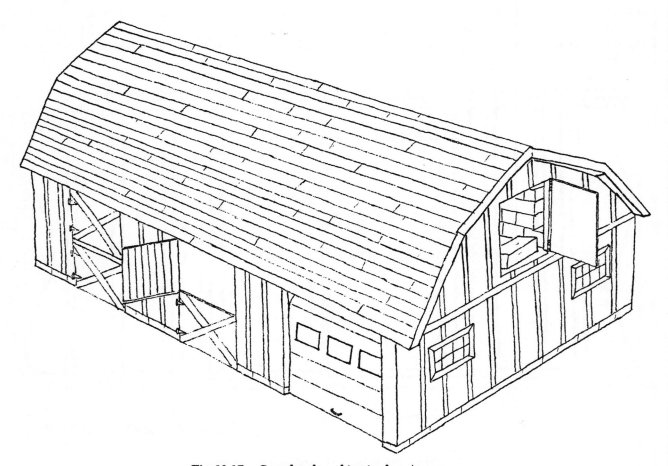

Fig. 10-17.   Completed combination barn/garage.

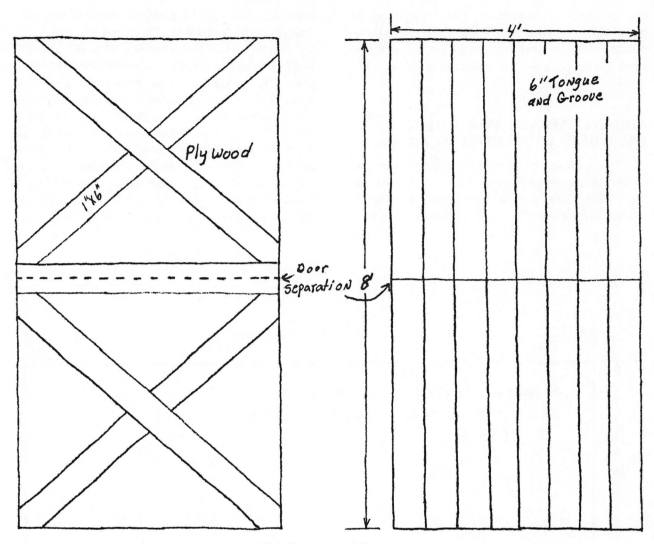

Fig. 10-18. Dutch door.

feed scale or even a bathroom scale to determine what size hinges to use. Use a **T** hinge. Your hardware should be able to tell you what size is needed. Of course, this has to be very strong to form a kick-proof door. No use having a heavy door if the horse can kick the hinge off. The door should be secured with a bolt lock and at night the bolt should be padlocked to prevent anyone from bothering the horses.

Goats also are becoming a valuable animal and they should be made secure if they will be bothered. Each one of these stalls can accommodate two goats

with all the paraphernalia needed to milk and care for them.

RUSTICATING THE EXTERIOR

No additional siding needs to be used for the exterior of the building, but 1 × 2 batten boards should be placed vertically on 16-inch centers to project a rustic appearance. The barn can then be painted red and trimmed with white. The very best way to finish the outside of this project, though, is to tear down an old barn and use the weathered boards as board and batten siding on your barn. Weathered boards should not be painted but they should be coated with three coats of a good waterproof preservative.

ARRANGEMENTS FOR DRYING PRODUCE AND SMOKING MEAT

The northwest corner of the building can be used for drying garden produce or for smoking meat (see Fig. 10-19). Make sure the room is well ventilated for at least two weeks after smoking meat before an attempt is made to dry vegetables, unless some smoke flavoring of the vegetables is not objectionable.

Meat can be smoked very well in this room by placing a plate of wood chips on the element of a hot plate. Drying heat can be generated by a small thermostatically controlled electric heater such as a "milk house heater." Provide a ventilator by extending a 4-inch diameter galvanized smoke pipe through the loft floor. By providing this pipe with a damper the amount of ventilation can be controlled. If meat or fish is soaked in salt water before being hung to smoke, a container should be placed to catch the salt water drippings if the room has a cement floor.

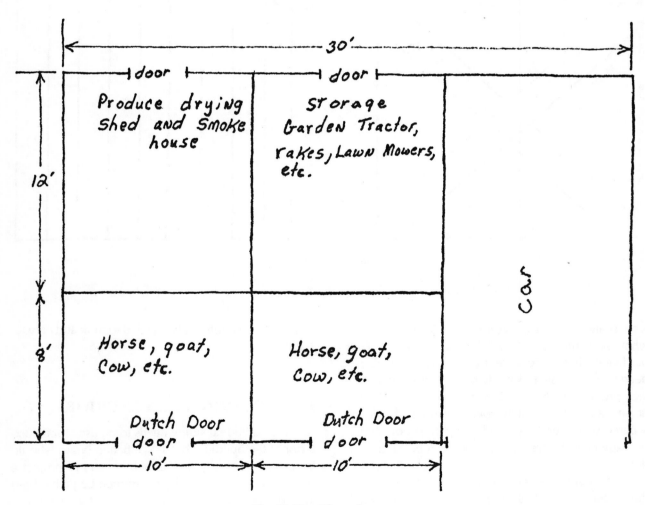

Fig. 10-19.   Floor plan.

Build drying racks in the room by covering frames made of 1 × 2 furring strips with ¼-inch mesh wire. The frames can be as wide as desired, and a different tier can be used for each foot of height. The same racks that will be used for drying produce can be used for smoking meat and fish also, but be sure that no salt water comes in contact with galvanized wire.

No special equipment is needed in the tool storage room except for a few sections of perforated boards nailed to the walls to hang tools from and perhaps a sheet of ¾-inch plywood lying on the floor to use as a platform for repairing and servicing garden tools. The garage stall can have an overhead snow tire platform as well as a lifting hoist for making repairs on the car.

The horse stalls or goat stalls should be built with a hay platform while the goat stalls should be made with a milking platform and other necessities of goat husbandry. Get a copy of *The Homesteader's Handbook* for complete information on the proper way to set up the stall and care for these animals.

Index